AF861810

Drip Irrigation for Row Crops

by
Blaine Hanson, Irrigation and Drainage Specialist
Larry Schwankl, Irrigation Specialist
Stephen Grattan, Plant-Water Relations Specialist
Terry Prichard, Water Management Specialist

University of California Irrigation Program
University of California, Davis
Revision I - 1997

Funded by the California Energy Commission and the U.S. Department of Agriculture Water Quality Initiative

To order or obtain ANR publications and other products, visit the ANR Communication Services online catalog at http://anrcatalog.ucanr.edu/ or phone 1-800-994-8849. You can also place orders by mail or FAX, or request a printed catalog of our products from

University of California
Agriculture and Natural Resources
Communication Services
1301 S. 46th Street
Building 478 - MC 3580
Richmond, CA 94804-4600
Telephone 1-800-994-8849
510-665-2195
FAX 510-665-3427
E-mail: anrcatalog@ucanr.edu

Publication 3376
ISBN-13: 978-1-60107-922-0

POD-12/15-SB/CR

Contents

Putting a System Together (continued)

Managing a Drip System

Fertilizing With A Drip System

Heading Off Problems

The Bottom Line

Preface

Drip Irrigation for Row Crops is the fifth in a series of water management handbooks prepared by the University of California Irrigation Program to help water managers address a wide range of practical irrigation matters. This handbook covers annual crops, while another handbook in the series, *Micro-irrigation of Trees and Vines,* covers permanent crops. Funding for the book was provided by the California Energy Commission and the U.S. Department of Agriculture Water Quality Initiative.

Other titles in this publication series include: *Agricultural Salinity and Drainage; Surge Irrigation; Irrigation Pumping Plants* and *Surface Irrigation.* Ordering information appears on the back of the title page. The authors thank Ricardo Amon of the California Energy Commission staff and Eric Zilbert, UC Agricultural Educational Specialist, for their help in making this publication possible, Ann Jackson for editing and preparing the original manuscript, and Tania Heta for preparing the revised manuscript.

Introduction

More and more growers in California are using drip irrigation on row crops. Drip irrigation — applying small amounts of water slowly and frequently through emitters spaced along polyethylene tape or tubing—is now the main method used in California to irrigate strawberries. Many lettuce, cauliflower, broccoli, celery, and tomato growers are also converting to drip irrigation.

The reasons for the growing popularity of drip irrigation are several. Drip irrigation offers improved yields, requires less water, decreases the cost of tillage, and reduces the amount of fertilizer and other chemicals to be applied to the crop. Because drip irrigation makes it possible to place water precisely where it is needed and to apply it with a high degree of uniformity, it lessens both surface runoff — excess water running off the lower end of the field—and deep percolation—water flowing down through the soil past the root zone where cannot be used by the crop. Drip irrigation can also be used in conditions unsuitable for other irrigation methods—on steep and undulating slopes, in very sandy soils, and in fields with widely varying soils.

These features make drip irrigation potentially much more efficient than other irrigation methods, which can translate to a significant water savings. But drip irrigation can only achieve this level of high efficiency if the system is carefully designed and managed so as to prevent such problems as emitter clogging and differences in emitter flowrates stemming from pressure variations in the irrigation system or from differences in emitters and flow passages originating in the manufacturing process.

This handbook is intended to serve as a practical guide to designing and managing a drip system for irrigating row crops and to addressing the unique problems associated with drip irrigation. It discusses patterns of root development, soil water content, and soil salinity under drip irrigation; how to apply water uniformly; how to determine how much water is being applied, how often to irrigate, and how long each irrigation should be; how to inject chemicals and fertilizers through a drip system; and how to prevent clogging from chemical precipitates, organic matter, and roots.

Some of the material used in this manual were obtained from manufacturer sources. Use of this material does not imply an endorsement of that product.

The information presented here is grounded both in technical literature, and in the authors' field experience. Comments or questions about the handbook or about drip irrigation in general can be directed to: Blaine Hanson, UC Irrigation and Drainage Specialist; Larry Schwankl, UC Irrigation Specialist; and Stephen Grattan, UC Plant-Water Relations Specialist, Department of Land, Air and Water Resources, University of California, Davis, CA 95616, (916) 752-1130; or Terry Prichard, UC Water Management Specialist, 410 S. Wilson Way, Stockton, CA 95205, (209) 468-2086.

Tables

Figures

What Does a Drip System Offer?

The Advantages and Disadvantages of Drip Irrigation

By Larry Schwankl, UC Irrigation Specialist

Like surface and sprinkler irrigation, drip irrigation offers both advantages and disadvantages. A carefully designed and managed drip system can make the most of these advantages and can often compensate for the disadvantages as well.

Advantages

• ***Improved plant response.*** With a drip system, irrigation takes place on a frequent basis (often daily), enabling the water manager to maintain the soil moisture at an optimum level. A well-designed, well-maintained drip system can also apply water more evenly than other irrigation methods. These features, together with effective irrigation scheduling, can improve plant growth and yield.

• ***Increased irrigation efficiency.*** Drip irrigation systems can make irrigation more efficient and therefore may require less applied water than sprinkler or surface irrigation systems — although these other systems can achieve similar levels of efficiency if the water manager can commit the time and resources necessary for optimum irrigation performance.

A drip system can improve irrigation efficiency by:

• reducing evaporation from the soil surface, since only a portion of the field surface is wetted;

• reducing or eliminating runoff;

• reducing deep percolation — water passing below the root zone;

• eliminating the need to drastically over-irrigate some parts of the field to compensate for the water being applied unevenly.

• ***Improved chemical application.*** Injecting fertilizers and other chemicals through a drip system has a number of advantages over conventional chemical application methods. When a drip system is used for chemigation or fertigation:

• the chemicals are applied to the soil in areas only where roots are actively growing;

• since irrigations are frequent, chemical applications can be better timed, making it possible to closely match fertilizer delivery with plant nutritional needs;

• because deep percolation is lessened, chemicals are less likely to be lost by moving past the root zone with the water. Potential harm to the environment is therefore reduced.

• ***Reduced weed growth.*** Since only a portion of the field is wetted, total weed growth may be less, although weed problems may increase in the areas kept continually wet. The continual wetting may accelerate the breakdown of herbicides used to control weeds.

• ***Decreased energy requirements.*** A drip irrigation system may require less energy than a conventional pressurized system if it increases irrigation efficiency and therefore requires less water to be pumped. Switching from a conventional pressurized system to a drip system can reduce energy needs by as much as 50%. But switching to a drip irrigation system from a surface irrigation system with water delivered through an irrigation district may *increase* energy requirements.

A properly designed and operated system allows the water manager to easily take advantage of lower energy rates offered during off-peak hours, but drip systems designed to operate 24 hours a day during peak crop water demand periods (which in any case is not recommended) would not be able to take advantage of off-peak rates.

• ***Automation.*** A drip irrigation system can be easily automated using electrical solenoid valves and a controller. This allows the system to be operated any time of the day or night and for any desired length of time, enabling water managers to take advantage of available crop water use information in determining optimum irrigation time. Soil moisture sensing devices such as tensiometers and electrical-resistance blocks can be used to further automate drip irrigation systems.

• ***Reduced salinity hazard.*** A drip irrigation system can lessen salinity problems by:

• diluting salts in the vicinity of the emitter by keeping soil moisture content at optimum levels through frequent irrigations;
• moving salts out to the edges of the wetted soil area, leaving a zone of wetted soil with a lowered salt content for root activity;
• eliminating the opportunity for salts to be absorbed through leaves, as may occur in sprinkler irrigation.

• ***Adaptability to difficult soil and terrain conditions.*** Drip irrigation can be used successfully on steep or undulating terrain not suited to any other irrigation system type, on soils with low water infiltration rates, and on soils with low water-holding capacities.

Disadvantages

• ***Maintenance Requirements.*** The small passageways characteristic of drip emitters make drip irrigation systems particularly susceptible to clogging from particulate matters, organic matter, roots, and chemical precipitates. Incorporating filtration in the system design can lessen clogging but additional maintenance—including injecting chlorine or acid and routinely flushing lateral lines — may be necessary to ensure top performance.

Machinery, animals, insects, or foot traffic in the field can cause leaks in the drip tape. Some growers make a practice of inspecting the system and correcting any problems before each irrigation, but visual inspection of subsurface drip systems is difficult.

• ***Cost.*** The underground PVC pipe, polyethylene tubing, filters, and other hardware required make drip systems among the most expensive of all irrigation systems. But the benefits — making very uniform water application possible and allowing the water manager to control the amount and timing of irrigations — may justify the expense. Growers contemplating drip irrigation should weigh the costs against benefits of a drip system according to the conditions at hand.

• ***Restricted root zone.*** Particularly in regions of low rainfall, plant root activity is often limited to the soil zone wetted by the drip emitters — usually a much smaller soil volume than that wetted by full coverage sprinkler or surface irrigation systems. While crops grown under drip systems have been shown to respond well, the drip system manager must remember that the system is meant to apply small, frequent irrigations — since, unlike crops grown under surface and sprinkler systems, crops grown under drip irrigation do not have large reserves of water stored in the soil and therefore cannot endure long periods between irrigations. Problems in the drip system — especially those occurring during peak water use — must therefore be repaired quickly to prevent water stress damage to the crop.

• ***Salt accumulation near the root zone.*** Unlike surface and sprinkler irrigation systems, which can flush salts below the crop root zone, surface drip systems tend to move salts to the outer edge of the wetted volume of soil and soil surface. In subsurface drip irrigation, salts will accumulate above the drip tape. Careful management is therefore necessary to ensure that the salts do not migrate back into the active root zone. Rainfall insufficient to totally leach the salts from the root zone can move the salts into the root zone and cause damage.

If the need to leach salts from the root zone becomes critical, a sprinkler or surface irrigation system may have to be used to accomplish this purpose effectively.

• ***Seed germination.*** A secondary irrigation method may be needed to germinate seeds.

• ***Drip tape disposal.*** After it is removed from a field, drip tape must be disposed of in landfills, creating additional costs and environmental problems. Where drip tape is replaced often, disposal could amount to a significant problem.

Applying Water Evenly and Efficiently

By Blaine Hanson, UC Irrigation and Drainage Specialist

The more evenly water is applied over a field, the potentially more efficient the irrigation. If every part of the field were to receive the same amount of water, uniformity would be 100 percent, but as a practical matter, no irrigation system can achieve 100 percent uniformity—some parts of the field always receive more water than other parts. When the uniformity of a system is particularly poor, some parts of the field may have to be severely over-irrigated so that the areas receiving less water in a given period will be adequately irrigated. This over-irrigating can cause excessive deep percolation.

Differences in Emitter Flowrates

In drip irrigation, uniformity depends on the extent to which emitter flowrates are the same throughout the field. The more variation in emitter flowrates, the less uniform the irrigation. Differences in emitter flowrates stem from the interplay of several factors:

• emitter clogging—the physical blocking of emitter flow passages as a result of materials suspended in the water, precipitated chemicals, slimes and bacteria growing within the system, or root intrusion;

• pressure variation within the irrigation system caused primarily by friction in mainlines, submains, and laterals and by elevation differences in the system;

• the relative sensitivity of the emitters to pressure differences;

• variations in the size or shape of emitter flow passages and orifices caused by eccentricities in the manufacturing process—called the *coefficient of manufacturing variation,* or *CV.*

• the system design, in particular, the length of the lateral lines, diameter of laterals, and characteristics of the emitters used.

Emission Uniformity

The uniformity of a drip system can be defined by calculating the *emission uniformity,* (sometimes referred to as the *distribution uniformity*)—specifically, the minimum emitter flowrate—usually the average of the lowest one-fourth of measured emitter flowrates—divided by the average emitter flowrate. These values are obtained by measuring emitter flowrates throughout the irrigation system.

Figures 1, 2, and 3 illustrate the relationship between emission uniformity and measured emitter flowrates along a drip lateral. The lower the emission uniformity, the greater the differences in emitter flowrates. Figure 1 shows that when emission uniformity is 95 percent, emitter flowrates vary only slightly. At an emission uniformity of 83 percent (Figure 2), differences in discharge rates are greater; at an emission uniformity of 69%(Figure 3), the differences are greater still.

In effect, the emission uniformity represents a measure of the potential for improving the irrigation system, as well as an estimate of the potential irrigation efficiency (the amount of irrigation beneficially used divided by the amount of water applied) for the adequately irrigated field — that is, a field in which the least-watered areas have received enough water for the crop.

If the average minimum amount of applied water equals the amount beneficially used, the emission uniformity becomes an estimate of the irrigation efficiency. The greater the emission uniformity, the greater the potential efficiency of an adequate irrigation.

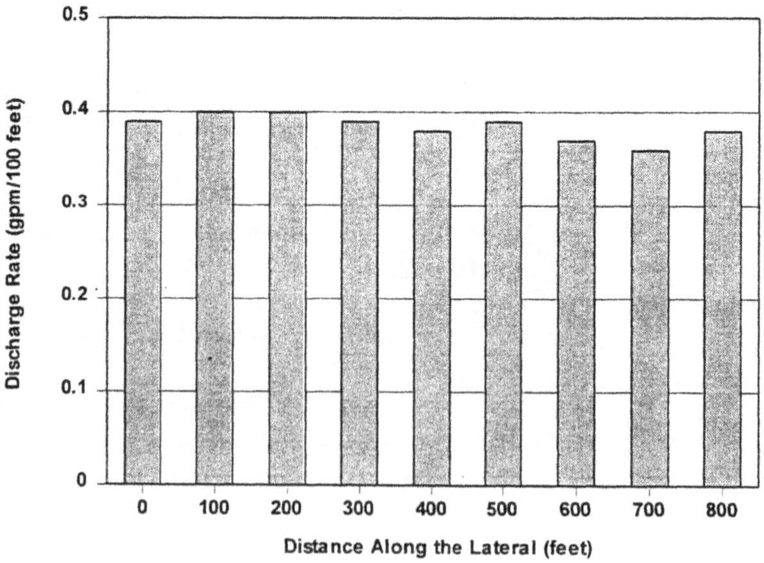

Figure 1. Relationship between emitter flowrates and a 95 percent emission uniformity.

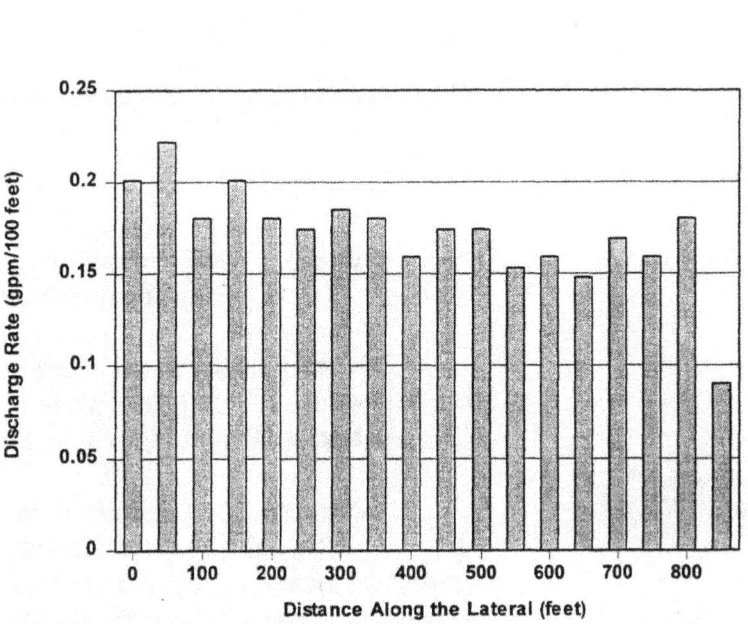

Figure 2. Relationship between emitter flowrates and an 83 percent emission uniformity.

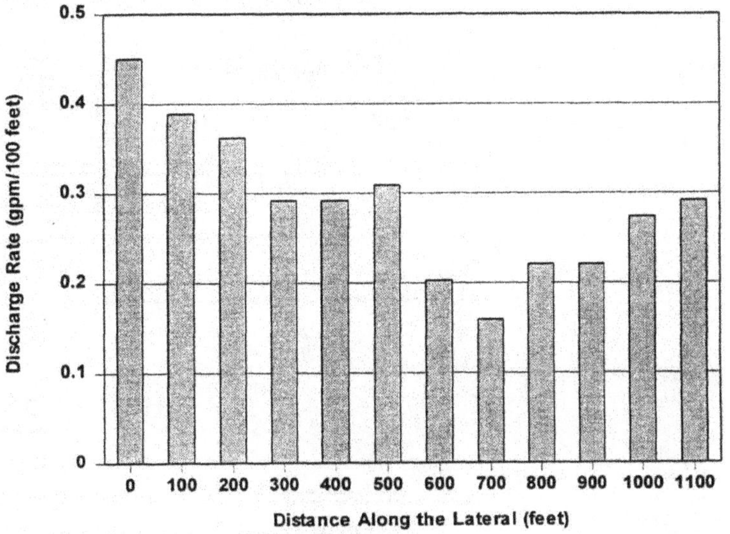

Figure 3. Relationship between emitter flowrates and an 69 percent emission uniformity.

Designing for Emission Uniformity

The following equation can be used to determine the emission uniformity of a particular drip system design:

$$EU = EU_{cv} \times EU_{pressure}$$

where EU = design emission uniformity,
EU_{cv} = emission uniformity due to the coefficient of manufacturing variation *(CV)*,
$EU_{pressure}$ = emission uniformity in response to pressure variation throughout the irrigation system.

The EU_{cv} can be calculated from the following:

$$EU_{cv} = 100\,(1 - 1.27 \times CV \div \sqrt{N})$$

where CV = coefficient of manufacturing variation,
N = number of emitters supplying water to each plant.

Practical field-wide design emission uniformities should range from 80 to 90 percent. To achieve a high field wide EU, the EU along the drip lateral must exceed 90 percent (see "Designing a Drip System").

Discharge Ratio

Because the EU is defined as the ratio of the minimum tape discharge rate to the average discharge rate, it does not reflect the actual discharge variability throughout the field. A better measure of the actual variability is the ratio of the tape minimum discharge rate to the maximum tape discharge rate, called the discharge ratio.

The discharge ratio can be calculated with the following equation:

$$\text{Discharge Ratio} = \frac{21.6}{(118.8 - EU)}$$

Table 1. EU versus discharge ratios.

EU	Discharge Ratio
95	0.91
90	0.75
80	0.56
70	0.44
60	0.37

While Table 1 shows that discharge ratio for several EUs. The above equation was developed using data from numerous evaluations of emitter discharge rates along laterals. Note that the data in Table 1 shows that for an EU of 90 percent, the discharge ratio is 0.75 indicating that the minimum discharge rate is 75 percent of the maximum discharge rate.

What Field Evaluations of Actual Drip Systems Show

Evaluation of micro-irrigation systems conducted by mobile laboratories revealed that most micro-irrigation systems have emission uniformities much less than 80 percent. Little correlations occurred between age of the systems and the emission uniformity indicating that new irrigation systems were designed below the desired EU. These results show that substantial room exists for improvement in both design and management of micro-irrigation systems.

Applying Chemicals Evenly

A drip system also makes it possible to apply chemicals through the irrigation system, although how uniformly the chemicals are applied depends directly on how evenly the water is applied. Figure 4 illustrates how the application uniformity of a mass of nitrate along a lateral in a buried drip system closely matches that of the applied water.

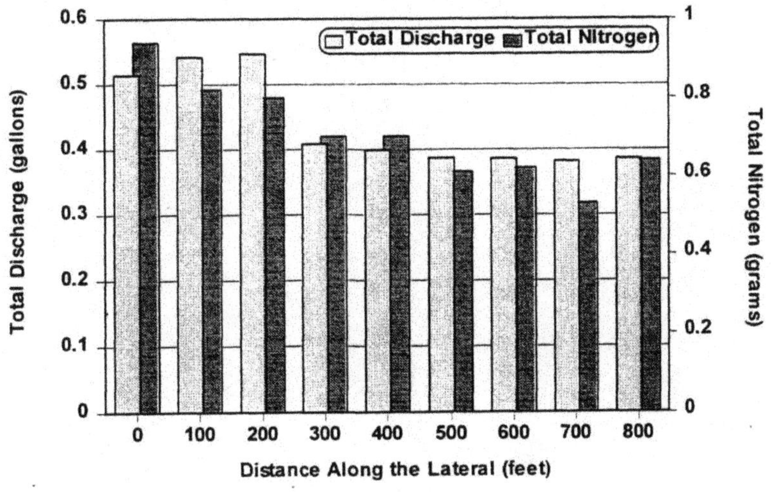

Figure 4. Application uniformity of water and nitrate along a lateral in a buried drip system.

Putting a System Together

Designing a Drip System

By Blaine Hanson, UC Irrigation and Drainage Specialist
and Larry Schwankl, UC Irrigation Specialist

Designing a drip irrigation system to suit a particular field and crop requires knowing how the individual components work, how the system operates as a whole, and about how the field conditions, the needs of the crop, and the quality of the irrigation water influence what kind of components are needed.

Components

Figure 1 on the next page shows the components of a typical drip irrigation system. Generally a drip system includes:

a pump
a flowmeter
mainlines, submains and manifolds
drip tape lateral lines
valves
a filter
injection equipment

Key Design Decisions

Designing a system requires making the following important decisions:

- what field flowrate and tape discharge rates are needed
- what kind of drip tape to use
- how long to make the lateral lengths
- what kind of pump to use
- whether the tape should be laid on the soil surface or buried underground—and if it is to be buried, how deep to bury it
- how far apart the emitters will be spaced
- whether the system will be used to germinate seeds or whether a backup system will be used for that purpose
- what kind of filtration and water treatment equipment is needed
- whether fertilizer and other chemicals will be injected through the drip system

Factors to Consider

Later chapters in this section provide detailed information to guide design decisions, but following is an overview of the main factors to be considered:

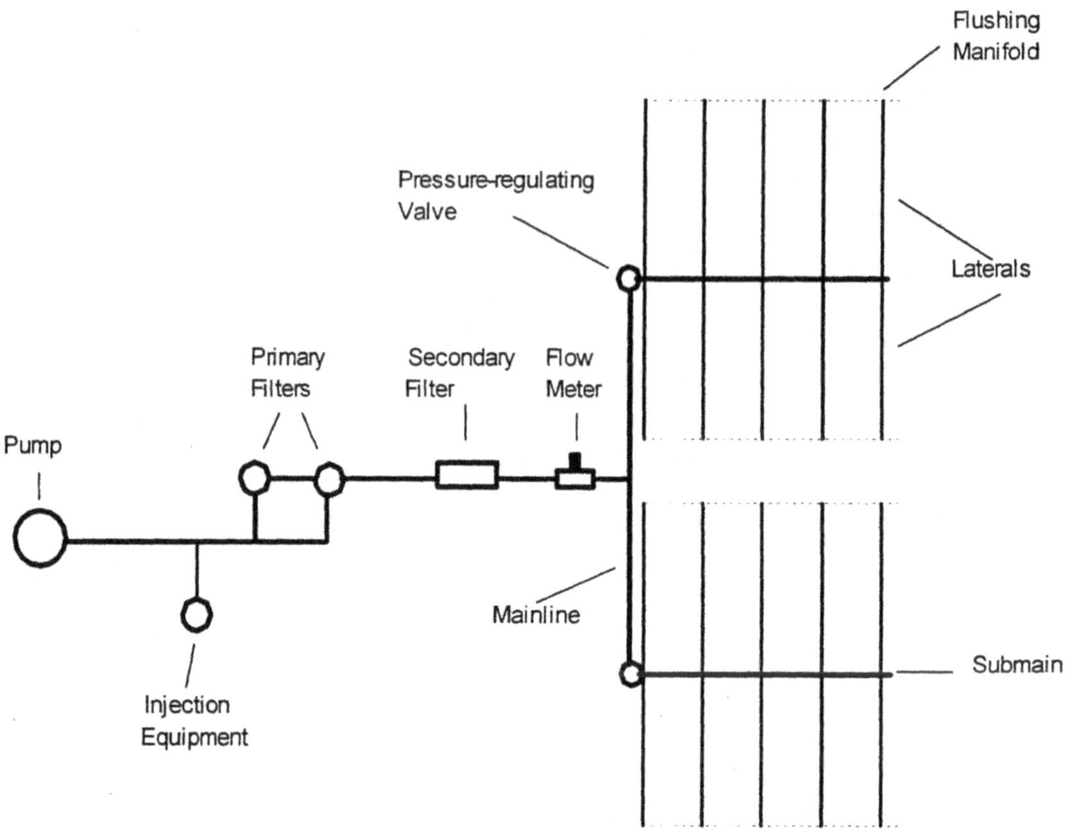

Figure 1. Components of a typical drip irrigation system.

Pumps

The pump and motor (or engine) chosen should be one that delivers the right pressure and flowrate as efficiently as possible. It is the responsibility of the drip system designer to determine what flowrate and pressure the pump should deliver, and it is the task of the pump dealer to use this information to designate the most efficient pump for a given system. For more information, see "Choosing a Pump".

Flowmeters

Flowmeters measure the volume of water moving through the system, making it possible to calculate how much water is being applied and therefore how often and how long the irrigation system should operate to be of the greatest possible benefit to the crop. Propeller flowmeters can report either the flowrate or the total volume of water applied. (See "Flowmeters: An Overview," and "Propeller Flowmeters".)

Mainlines and Submains

Main and submain pipes, usually made of PVC, deliver water to the lateral lines and emitters. Deciding what size pipe to use requires balancing the cost of the pipe against cost of energy caused by the effect of friction as water moves through the system, causing a gradual loss of pressure. A qualified drip irrigation system designer will take these factors into account in recommending pipe sizes. (See Appendix B "Calculating Friction Losses".)

Drip Tape

The guiding factor in choosing what kind of drip tape to use is the goal of making sure that water is applied over the field as evenly as possible. The discharge rates of the emitters in high quality tape are more uniform than the discharge rates of the emitters in lower quality tape. The *coefficient of manufacturing variation* — a figure that the manufacturer should supply — defines the variation in emitter discharge rates due to inconsistencies inherent in the manufacturing process. (See "Drip Tape and Tubing for Row Crops," and "Selecting a Drip Tape".)

The diameter of the tape and the length of the lateral lines are also part of the uniformity equation, as is the extent to which the system can adjust to changes in water pressure. If the lateral lines are too long or the wrong diameter, the emitters may discharge water at very different rates, making irrigation uneven. Pressure-compensating ability is particularly important if lateral lines are longer than 600 feet or if the field has steep up and down slopes. Incorporating pressure-compensating regulators or pressure-compensating drip tubing in the system can help overcome pressure differences and variation in emitter discharge rates to make irrigation more uniform. (See "Selecting an Appropriate Lateral Length" and "Pressure-Compensating Emitters".)

How long to make the lateral lines, what diameter tape to use, the effect of those decisions on the emitter discharge uniformity, and the cost of the tape, must all be balanced against one another in designing the system.

Drip tapes usually have a wall thickness of between 4 and 10 mil, although thicker tape walls of 15 mil have been found to reduce damage from ants and other insects.

Drip system design often assumes a tape life of two to three years, although tape can perform effectively for as long as ten years if carefully maintained to minimize clogging.

Valves and Regulators

Valves and regulators control water flow and pressure in the drip system. Check valves are installed at the pump discharge to prevent water from flowing backward into the well. Pressure-regulating valves help maintain a constant downstream pressure. Pressure-relief valves protect against pressure surges that might damage pipelines. Vacuum-relief valves installed either at the submain or at the flushing manifold prevent a vacuum from forming inside the tape, which could cause soil to be sucked into the emitter and the emitter to become clogged. (See "Valves and Regulators" and "Preventing Backflow".)

How Deep to Bury Subsurface Drip Tape

If the drip tape is to be buried, the usual burial depth is between 6 and 12 inches — rarely more than 12 inches, except for research purposes, although in root crops such as sugar beets and potatoes, deeper placement may be desirable. Soil type, crop type and root pattern, soil salinity, soil surface wetting patterns, and tilling practices should all be taken into account in deciding how deep to bury the tape. (See "Deciding How Deep to Bury Tape".)

Emitter Opening Orientation

Drip tapes should be installed so that the emitter openings face upward to prevent silts, clay, and other materials that may have settled out in the drip tape from entering and clogging the emitter.

Emitter Spacing

In the drip tapes currently available emitters may be spaced from a few inches apart to more than three feet apart, although usual spacing is between 12 and 18 inches. (See "Drip Tape and Tubing for Row Crops".)

Irrigation Water Quality

The irrigation water to be used should be analyzed to help determine what filters and chemical treatments will be needed to keep the emitters from clogging. Sand, silt, and algae in the water can block emitter flow passages, and calcium, magnesium, bicarbonate, iron, manganese, and sulfide can precipitate to also clog emitter flow passages. High concentrations of iron and manganese can promote the growth of iron slimes and bacteria, which can clog emitters. Fertilizers can also react with irrigation water in chemical reactions that may result in clogged emitters. A chemical analysis of the water can assess the potential for these reactions. (See "Assessing Water Quality" and "Water-Fertilizer Interactions".)

Filters

Vortex filters ("sand separators") can remove sand and other particulate matter suspended in the water. Screen filters are also effective in removing particulate matter. Screen or sand media filters are usually used for removing particulate matter finer than sand particles. Algae, slim, and other organic matter can be removed using either screen or sand media filters, but since organic matter can quickly clog a screen filter and is difficult to flush from the screen, sand media filters are the usual choice for filtering surface waters containing algae and slime. Disk filters are sometimes used in lieu of sand media filters. (See chapters on filters.)

Chemical Treatment

Chemical treatments remove biological micro-organisms such as algae, fungi, and bacteria and can remove chemical precipitates or prevent calcium, bicarbonate, iron, manganese, and sulfide from forming precipitates. (See section "Heading Off Problems".)

Soil Ingestion

After the water is turned off, the saturated soil adjacent to the emitter can prevent air from readily entering the laterals during drainage, causing a vacuum to develop in the lateral. The vacuum tends to suck soil particles into the lateral—a process called soil ingestion—which can clog the emitters as the ingested soil settles along the lower part of the lateral. Soil ingestion can be minimized by installing vacuum relief valves at the highest elevations of the lateral. (See "Soil Ingestion".)

Flushing

While filtration will remove large sand and other particles, silt and clay will move through the filter and may eventually settle in the laterals, usually where the water velocity is slowest. Soil ingestion during drainage can also cause material to settle in the laterals. Periodic flushing at a minimum flowrate of about 1 gallon per minute at the end of the lateral for a 5/8" diameter tape is suggested to remove any material that has accumulated in the laterals. A flushing manifold can be installed at the end of the laterals for this purpose, thus tying the drip lines together. (See "Flushing".)

Germinating Seeds

If a buried drip system is to be used for germinating seeds, the water must flow upward to the seeds from the tape. Studies show that seeds can germinate successfully if the drip tape is buried not more than 8 inches deep and if emitters are spaced about 12 inches apart, providing the seed bed is adequate. Soil type, rainfall, and other conditions specific to the site should be taken into account, however. A sprinkler irrigation system may be needed to germinate seeds under certain conditions. (See "Germinating Seeds".)

Injection Equipment

Drip irrigation is well suited to injecting chemicals such as chlorine and fertilizer. Various types of injection equipment can be used, including differential pressure tanks ("batch tanks"), venturi devices, and electrically driven or water-driven pumps, depending on the chemical applied, the accuracy level needed, and the injection rate required. (See "Injection Devices".)

Cost

The initial cost of a drip irrigation system is approximately $900 to $1300 an acre. (See "Will Using a Drip System Increase Profits" and "Will Converting to Drip Irrigation Save Energy?".)

Drip Tape and Tubing for Row Crops

By Blaine Hanson, UC Irrigation and Drainage Specialist

The types of drip tape and tubing used in row crops include: double-chambered tape, single-chambered tape or tubing, and porous tubing.

Double-Chambered Tape

Double-chambered tape (Figure 1) consists of a large main chamber—which conveys the water along the lateral—and a small secondary chamber. Widely spaced orifices installed between the main and secondary chambers discharge water into the secondary chamber. Orifices in the outer wall of the secondary chamber discharge water into the soil. This design is not used any more.

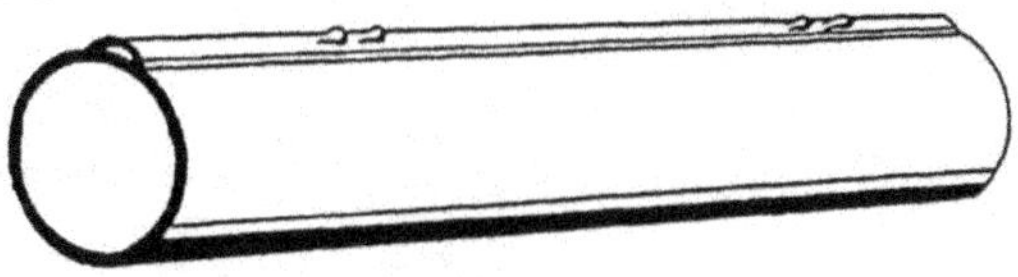

Figure 1. Double-chambered drip tape.

Single-Chambered Tape/Tubing

Single-chambered tape or tubing consists of a main chamber only. Water flows from this chamber into drip emitters spaced along the tape or tubing and is discharged from the emitters into the soil.

Porous Tubing

Porous tubing, sometimes made from recycled tires, has walls with pores of capillary size, through which water under pressure can seep.

Types of Emitters

Several different types of emitters can be used with single-chambered tape or tubing. One type commonly used is a strip emitter, which has a long, narrow shape (Figure 2). Most of these emitters are classified as turbulent-flow emitters because the configuration of the flow path creates turbulence in the water. The turbulent-flow configuration allows for larger flow passages, which in turn can reduce the potential for clogging.

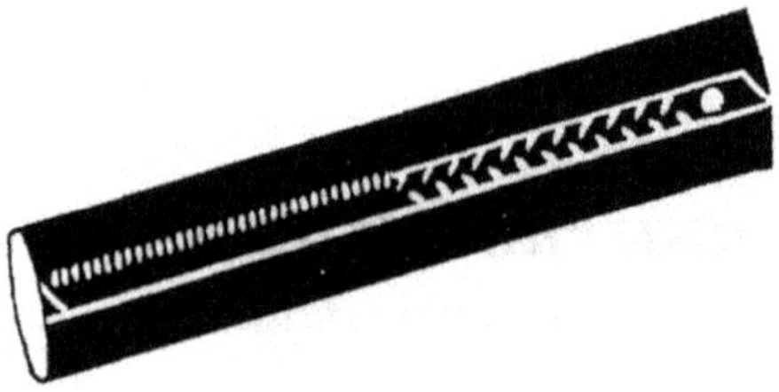

Figure 2. Strip emitter.
(T-Systems International)

Another emitter choice for drip tapes are individual emitters—usually turbulent-flow emitters—attached to the inside of the tape (Figure 3).

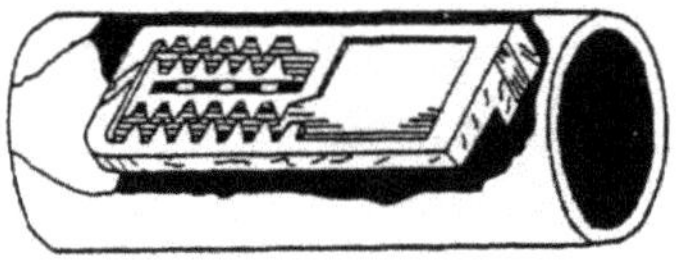

Figure 3. Welded emitter.
(Netafim Irrigation Company)

Still another emitter type available is the in-line emitter (Figure 4), which is installed inside the drip tubing as part of the tubing flow path.

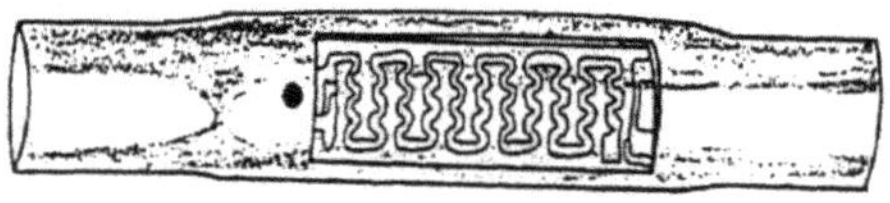

Figure 4. In-line emitter.

Pressure-compensating Emitters

Drip tubing with pressure-compensating emitters is also available. These emitters have flexible diaphragms or orifices that adjust the flow passage dimensions according to the water pressure. A minimum operating pressure (usually 4 to 10 psi, depending on the manufacturer) is required for the pressure-compensating feature to operate. Pressure-compensating drip tape or tubing is particularly appropriate in fields with significant elevation changes or where very long lateral lengths are used.

Tape Diameter

The diameter of most drip tapes and tubing is about 0.625 inches, although larger-diameter tape (0.875 inches) is available. Porous tubing and drip tape with a diameter of about 3/8 inches are also available.

Emitter Spacing

Emitter spacing varies considerably, depending on the manufacturer. The smallest spacings range from 2 inches to 4.5 inches. The largest spacing for most tapes is 24 inches, although larger spacings are available in drip tubing. For row crops, spacings usually range between 8 and 18 inches.

Tape Wall Thickness

Available wall thicknesses of drip tape range from 4 to 25 mils (one mil = 1/1000 of an inch). Thin-walled material installed on the surface is frequently replaced after each crop season. Drip tape with wall thicknesses of between 8 and 15 mils are often used for subsurface drip systems.

Operating Pressure

The recommended maximum continuous operating pressure for drip tape or tubing depends on the wall thickness of the tape. Several manufacturers recommend a maximum continuous pressure of about 15 psi for 15 mil tape and 8 to 12 psi for lesser thicknesses. Much higher pressures (60 psi) can be used with tubing.

Tape Discharge Rate

The discharge rate of drip tape is often expressed in gallons per minute per 100 feet of tape length This tape discharge rate is affected by the relationship between pressure and the emitter discharge rate and by the emitter spacing. For a given pressure/emitter discharge rate relationship, the closer the emitter spacing, the greater the tape discharge rate; and the wider the emitter spacing, the lower the tape discharge rate.

Some tape manufacturers provide a low-flow emitter (about 0.2 to 0.3 gallons per minute per 100 feet at a given emitter spacing), a high-flow emitter (greater than about 0.4 gallons per minute per 100 feet) and an ultra-high flow tape (more than 0.6 gallons per minute per 100 feet).

Chemicals

Also available are drip tapes that contain chemicals to lessen or prevent clogging. One such tape contains a strip of copper to reduce clogging from algae, bacteria, or other material; a drip tubing is impregnated with an herbicide to combat root intrusion.

Selecting a Drip Tape

By Blaine Hanson, UC Irrigation and Drainage Specialist

Every drip tape performs differently, with some discharging water more evenly than others. Figure 1, showing discharge rates along laterals of four different drip tapes, illustrates the differences in drip tape performance.

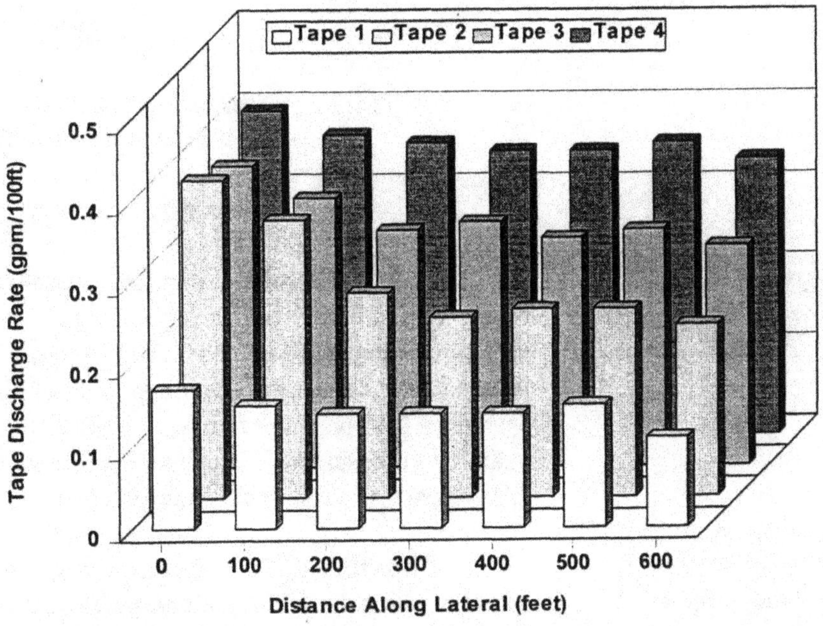

Figure 1. Discharge rates of four drip tapes.

Factors Affecting Uniformity

The uniformity of the discharge is affected by such factors as:

- the coefficient of manufacturing variation (CV) — that is, differences in output among the individual emitters in the system as a result of the manufacturing process
- the sensitivity of the discharge rate to pressure variation caused by friction in the tape or elevation differences in the field
- the tape's susceptibility to clogging
- the tape wall thickness and strength

Deciding which tape to buy requires balancing these factors against cost and the level of emission uniformity desired.

Coefficient of Manufacturing Variation (CV)

The CV for a given system — which the manufacturer should provide — is calculated by measuring the discharge rates of a sampling of emitters at a constant pressure. A CV of 0.05 or less is considered excellent; 0.05-0.10 is acceptable; and more than 0.1 is marginal. Most drip tapes have CVs of less than 0.05.

Relationship Between Emitter Discharge Rate and Pressure

The emitter discharge rate is affected by changes in water pressure, with some emitters more sensitive to pressure changes than others. This sensitivity to pressure changes is described by the *emitter discharge exponent.* The following equation describes the relationship between pressure and emitter discharge rate:

$$Q = (K)(P^x) \qquad (1)$$

where Q = emitter discharge rate
K = emitter discharge coefficient
P = pressure
x = emitter discharge exponent

Emitter Discharge Exponent

The smaller the exponent, the less sensitive the emitter discharge rate to pressure variation. An exponent of 0 means the emitter is completely able to compensate for pressure changes. These emitters are classified as pressure-compensating. An exponent of 1 means the emitter cannot compensate at all for pressure changes. With an exponent of 0.5, a 20 percent change in pressure will cause a 10 percent change in emitter flowrate. Most drip tapes have emitter discharge exponents of between 0.4 and 0.6.

Emitter Discharge Coefficient

The effect of the dimensions and configurations of flow passages is described by the *emitter discharge coefficient.* Large coefficients generally suggest large flow passages, although two different emitters may have the same coefficient but different flow passage dimensions because of turbulence in the water caused by flow passage configuration. The value of the discharge coefficient also depends on pressure units and discharge rate units.

Clogging

The relative sensitivity of a particular emitter to clogging is influenced by the dimensions and configurations of the flow passages and by any consequent turbulence in the water flowing through the emitter. Generally, large passages are associated with less potential for clogging, but actual clogging tendency should be determined from field experience and from manufacturer's recommendations.

Drip Tape/Tubing Characteristics

Distinct from the *emitter* performance characteristics, the characteristics of the *drip tape or tubing* should also be considered in choosing tape. Drip tape discharge rates, expressed in gallons per minute per 100 feet, depend on both the emitter discharge rate and the emitter spacing. The greater the spacing between emitters, the lower the drip tape discharge rate for a given emitter.

Emitter Spacing

The best emitter spacing to use may depend on how deep the tape is buried and on the depth of the crop root zone, although little research has been published on this question. Available emitter spacings range from several inches to several feet. An emitter spacing of twelve inches is often used for shallow installations in shallow-rooted crops, but larger spacings may be more appropriate for deep-rooted crops with deep installations. For deep installations and deep-rooted crops, emitter spacings up to about 40 inches have been used successfully, although this requires a second irrigation method to establish the crop.

High Flow/Low Flow

High flow drip tape is roughly defined as that having a tape discharge rate of 0.4 gpm or greater per 100 feet. A low flow tape is defined as that having a discharge rate of about 0.2 to 0.3 gpm per 100 feet for the same emitter spacing. An advantage of the high flow tape is that the larger flow passages of the emitter can reduce clogging problems.

Application Rate

The tape chosen for a drip system should be capable of providing the necessary application rate. The following equation can be used to calculate the application rate needed:

Inches per hour = 11.55 x gallons per minute per 100 feet ÷ furrow spacing (inches).

Tape Wall Thickness

Tape wall thicknesses range from 4 mil to 25 mil. The thicker the wall, the higher the allowable pressure. Thicker-walled tape may also lessen insect damage.

Tape Diameter

The diameter of most drip tape is 5/8" (0.625"). Several manufacturers offer a drip tape with a 7/8" (0.875") diameter, while one manufacturer offers a 3/8" (0.375") tape diameter. Large diameter tape is more expensive than the smaller diameter tape, but has the advantage that less pressure is lost in the system as the result of friction.

Root Intrusion

The susceptibility of the tape to root intrusion should be considered in designing a system. The emitter outlet should be located away from the seam if possible, since roots tend to grow along the seam. The emitters of one particular product are impregnated with a chemical (Trifluralin) to prevent root intrusion.

Which Drip Tape Is Best?

Which drip tape is best suited to a given situation depends on such factors as conditions at the site, type of crop, the cost of the tape, and the characteristics of the particular drip tape. The following should be considered:

- Where high cash value crops are grown on land with steep slopes or severe undulations, which could cause uneven emitter flowrates and significant variations in crop yield, a tubing with pressure-compensating capability might be the best choice. The increased crop yield from the improved uniformity and the resulting higher profit might completely offset the higher cost of the pressure-compensating material.

• The tradeoffs between the manufacturing CV and the emitter exponent should be taken into account. A tape with a relatively low exponent and a high CV may have better emission uniformity than a tape with a higher exponent and a low CV. Some guidelines for considering these tradeoffs are:

• Systems with short laterals lengths (less than 400 feet) on small downhill slopes or on level fields may have only slight pressure variation. In these conditions, the emitter discharge exponent is not critical and the tape selected should be one with a low CV.

• Systems with longer lateral lengths on fields with steep or undulating slopes or uphill slopes may have significant pressure variation. In these conditions, the emitter discharge exponent should be the first consideration in selecting a tape, since a low exponent will reduce the variability in emitter discharge from pressure differences. The CV should be the second factor considered.

• It may be possible to reduce the capital and installation costs of submains, manifolds, and other components by choosing pressure-compensating tape or tubing or by using tape with a large diameter.

Pressure-compensating features may allow longer lateral lengths, perhaps making it possible to eliminate submains and manifolds, but the longer lateral lengths may also increase energy costs. A larger diameter tape will reduce pressure variation.

Table 1 on the next page shows the characteristics of the drip tapes used in Figure 1. Tape 1 was a low-flow tape (5/8" diameter), resulting in only slight pressure losses from friction. The discharge exponent showed that the tape was sensitive to pressure changes, but the tape discharge rate varied only slightly because of the minimal pressure loss.

Tape 2 was a high-flow tape (5/8" diameter) with a discharge exponent lower than that of Tape 1, indicating that the discharge rate was less sensitive to pressure changes. But the high flowrate of this tape caused significant losses in pressure from friction, resulting in more variability in discharge rates. This tape had the lowest emission uniformity.

Tape 3, also a 5/8" diameter tape, had a low discharge exponent (0.29), indicating the tape discharge rate was not very sensitive to pressure differences. Emission uniformity was high with this tape, even though the CV was higher than that of Tape 3.

Tape 4 had a discharge exponent similar to that of Tape 2, but the tape diameter was 7/8". The larger tape diameter resulted in less pressure loss and a higher emission uniformity than the other tapes, but Tape 4 cost about $0.046 per foot compared to $0.031 per foot for Tapes 1 and 2 and $0.020 per foot for Tape 3. The wall thickness of Tape 2 was 9 mil, while the wall thickness of the other three tapes was 15 mil, which probably accounts for the cost difference between Tape 2 and Tape 3.

Table 1. Performance characteristics of drip tapes shown in Figure 1.

	CV	Exponent	EU (%)
Tape 1	0.073	0.86	87
Tape 2	0.026	0.69	81
Tape 3	0.072	0.29	91
Tape 4	0.020	0.61	96

Fortunately, growers have many high quality drip tape products from which to choose. Careful selection, design, and maintenance should result in an drip irrigation system with high uniformity.

Coefficients of manufacturing variation and emitter discharge exponents for various drip tapes and tubings can be obtained by writing or calling the Department of Land, Air and Water Resources, University of California, Davis, California 95616, (916) 752-1130.

Design Criteria and System Layout

By Blaine Hanson, UC Irrigation and Drainage Specialist

Introduction

One advantage of drip irrigation is the potential to apply water very uniformly throughout the field. This potential will be achieved, however, only if the system is properly designed with appropriate lateral and manifold lengths and both adequate pressure and pressure regulation throughout the irrigation system. Field evaluations of micro-irrigation systems have shown that most micro-irrigation systems (drip tape, drip emitters, and microsprinklers) are not achieving this potential.

System Layout

Figure 1 shows a typical design of a drip irrigation system. The field is split into blocks, each block containing lateral lines connected to a manifold. The manifold, in turn, is connected to a submain. Some type of pressure control is installed at the manifold inlet. Submains are then connected to the mainline. A flushing manifold connected to the end of the laterals may be installed at the end of the block.

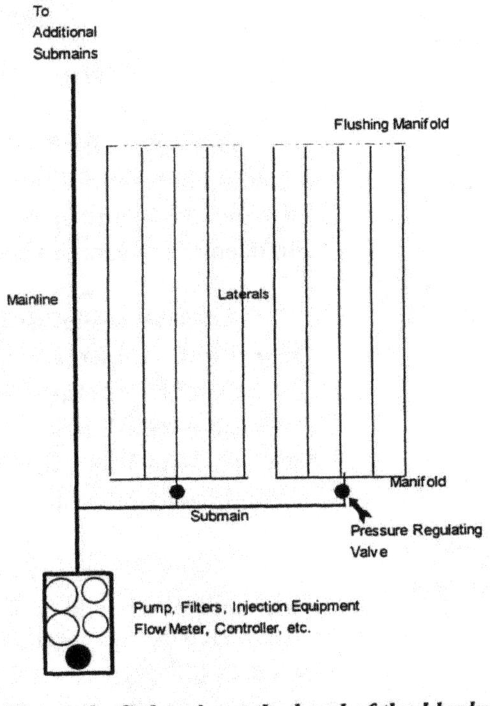

Figure 1. Submain at the head of the blocks.

Several other design approaches can be used. A second submain and manifolds, installed at the end of a block, supplies water to both ends of the laterals

(Figure 2). This design allows longer lateral lengths, more conducive to cultural practices. One of the supply manifolds is used as a flushing manifold. A disadvantage of this approach is that the pressure needed for flushing flowrates may exceed the manufacturer's recommended maximum pressure.

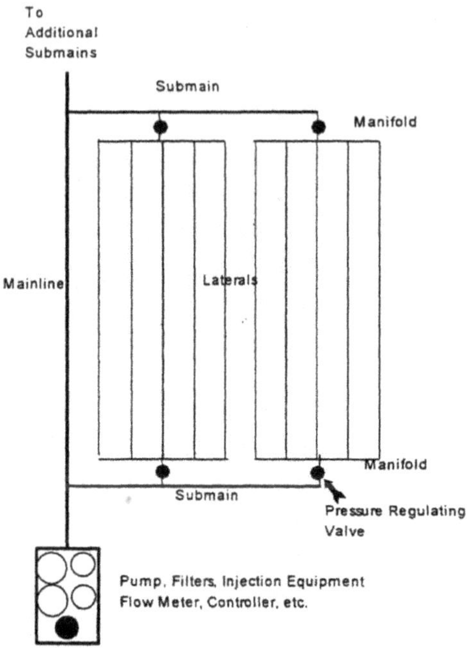

Figure 2. Submains at both ends of the blocks.

Another variation, appropriate for fields with small slopes, is to install the submains and manifolds in the middle of the block, with laterals on both sides of the manifold (Figure 3). This also allows longer field lengths. Flushing manifolds will need to be installed at both ends of the block.

A design approach for fields with steep slopes is to offset the submains and manifolds from the middle of the field (Figure 4). Smaller lateral lengths will occur upslope from the submain and longer lateral lengths will occur downhill. One design investigation for a field 1200 feet long with a 1% slope showed the highest system uniformity with the submain and laterals installed 500 feet from the uphill end of the field.

Design Uniformity

The emission uniformity (EU) is an index used to describe how uniform water is applied. The emission uniformity is defined as the minimum tape or emitter discharge rate divided by the average discharge rate. The distribution uniformity (DU) is also used as a uniformity index. Its definition is the minimum infiltration divided by the average infiltration. The EU or DU of the irrigation system is the potential maximum irrigation efficiency of a properly-managed irrigation system. Thus, the higher the EU, the higher the potential maximum irrigation efficiency.

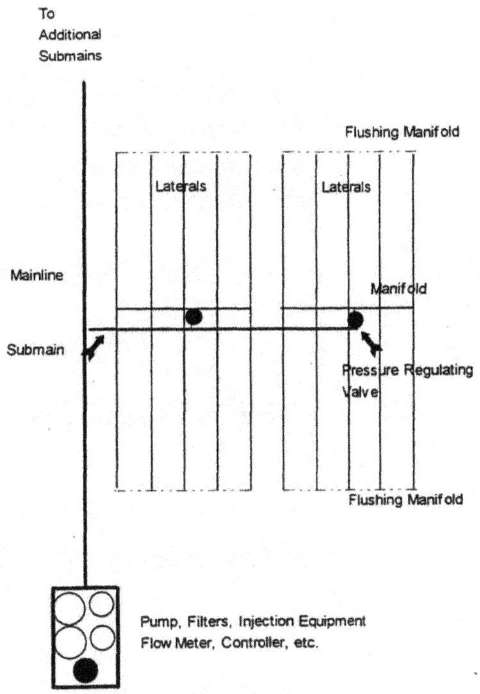

Figure 3. Submain in the middle of the blocks.

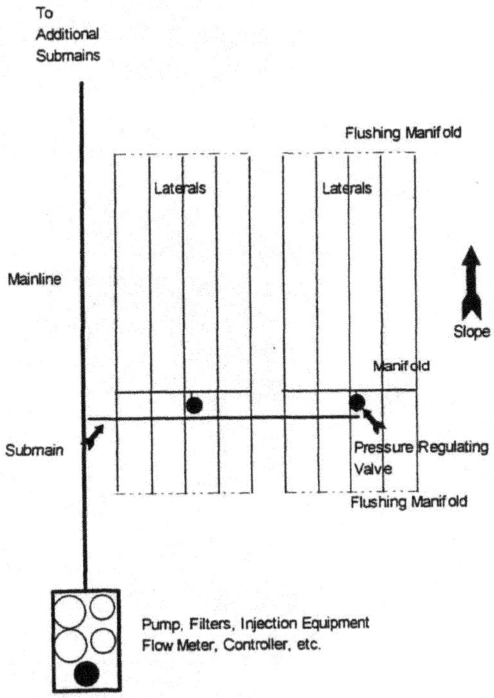

Figure 4. Submain offset from the middle of the blocks.

The EU or DU is normally used as the design criterion for uniformity. However, field experience indicates that excessive soil wetness can occur near the head of a lateral, even at EU's considered to be acceptable. The excessive wetness causes trafficability problems and/or crop growth problems, The EU as an indicator of differences between maximum and minimum discharge rates is inadequate, since the EU is the ratio of the minimum discharge rate over the average discharge rate. Thus, the design criteria for laterals should be based on both the ratio of minimum to maximum discharge rates and the EU.

The field-wide or irrigation system emission uniformity is described by:

$$EU_{SYSTEM} = EU_{BETWEEN\ BLOCKS} \times EU_{WITHIN\ BLOCK}$$

The emission uniformity within the block is:

$$EU_{WITHIN\ BLOCK} = EU_{LATERAL} \times EU_{MANIFOLD\ PRESSURE}$$

and the emission uniformity of the lateral is:

$$EU_{LATERAL} = EU_{CV} \times EU_{LATERAL\ PRESSURE}$$

where EU_{CV} is the emission uniformity caused by manufacturing variation.

Because the system EU is the product of the individual component EU's, the component EU's must be high to obtain a reasonable EU_{SYSTEM}. For example, if the EU_{CV} = 97%, $EU_{LATERAL\ PRESSURE}$ = 90%, $EU_{MANIFOLD\ PRESSURE}$ = 97%, and $EU_{BETWEEN\ BLOCKS}$ = 96%, then the system EU is:

$$EU_{SYSTEM} = 100 \times 0.97 \times 0.90 \times 0.97 \times 0.96 = 81\%.$$

A key to achieving a high system EU is to design a lateral with a very high EU, which involves both minimizing emitter discharge-rate variability due to pressure changes along the lateral length and selecting a drip tape or emitter with a low coefficient of manufacturing variation (CV). Based on field evaluations of micro-irrigation systems, the following design EU standards are recommended:

$$EU_{LATERAL} = 90\% \text{ or greater,}$$
$$EU_{SYSTEM} = 80\% \text{ to } 90\%.$$

Obtaining an $EU_{LATERAL}$ greater than 90% means designing a lateral such that the $EU_{LATERAL\ PRESSURE}$ is about 95%.

The following equation can be used to determine the ratio of minimum to maximum tape discharge rates, or the discharge ratio for a given emission uniformity:

$$Q_{MINIMUM} / Q_{MAXIMUM} = 21.6/(118.8 - EU)$$

A design method, "Designing Laterals for Drip Irrigation Systems", uses this relationship for designing laterals of drip irrigation systems. This procedure is in Appendix C.

Which design option is the least costly? An analysis of different design options for a 40 acre and 160 acre field showed the results shown in Table 1.

Table 1. Cost analysis for drip irrigation systems. Crop was tomatoes with a bed spacing of 5 feet.

Location of Submains and Manifolds	Drip Tape	Lateral Length	Relative Cost (%)
40 Acres			
Head of block	High-flow, 5/8"	600	100
Middle of block	High-flow, 5/8"	600	105
Head of block	Low-flow, 7/8"	1300	88
160 Acres			
Head of field	High-flow, 5/8	600	100
Head of field	Low-flow, 5/8	850	72
Head of field	Low-flow, 7/8	1300	64

For both field sizes, the design using 7/8-inch diameter drip tape cost the least. The longer the lateral length, the smaller the system cost.

Sources of Nonuniformity

Sources of nonuniform water application include variability in emitter or tape discharge rates within a block and variability between blocks. Variability within a block depends in changes in the discharge rates along the lateral length and changes along the manifold length. Variability along a lateral length, in turn, depends on lateral inflow rate, lateral length, diameter of the drip tape, slope, emitter spacing, and emitter discharge rate characteristics, coefficient of manufacturing variation. Some of these factors are interrelated. For example, the lateral inflow rate will depend on lateral length, emitter spacing, emitter discharge rate characteristics and diameter of tape.

Variability in tape discharge rates within a block can be reduced by the following measures:

1. Use low-flow drip tape.

2. Increase emitter spacing to a recommended maximum spacing of 16 to 18 inches.

3. Use appropriate lateral and manifold lengths.

4. Use a larger diameter drip tape or manifold tubing.

5. Use a drip tape with a smaller emitter discharge exponent. The relationship

between the tape or emitter discharge rate (q) and pressure (P) is;

$$q = (K)(P^X)$$

where K is the discharge coefficient and X is the discharge exponent. The lower the discharge exponent, the less sensitive the tape or emitter discharge rate to pressure changes. (See "Selecting a Drip Tape".)

Variability between blocks depends on the pressure variation along submains and mainline and on the use of pressure regulating valves. If pressure-regulating valves are used, then variability between blocks will depend on the quality of the pressure regulation.

Selecting an Appropriate Lateral Length

By Blaine Hanson, UC Irrigation and Drainage Specialist

Irrigating efficiently by applying water as evenly as possible is achieved in drip irrigation by minimizing differences in emitter discharge rates throughout the field. Selecting the appropriate lateral length is the key to minimizing these differences. The design emission uniformity of the drip lateral should be at least 90 percent to achieve a field-wide emission uniformity between 80 and 90 percent.

The Effect of Friction on Pressure

Friction from the water flowing through the drip tape causes pressure to decline along the lateral. The diameter of the tape, the roughness of the tape material, the flowrate in the tape, and the length of the lateral all determine how much friction is generated in the system and therefore the extent to which pressure varies as water flows through the system components. A commonly-used equation for calculating friction losses is in Appendix B.

Figure 1 illustrates the effect of friction on pressure and tape discharge rate along a lateral where the slope is 0%. For this case, friction is the only force acting to reduce pressure. Pressure is highest at the inlet and decreases rapidly along the upper part of the lateral, then more slowly along the lower part of the lateral. The rapid decrease is caused by friction losses from the high flowrates in the drip tape along the upper part of the lateral, but as water is discharged through the emitters, the flowrate along the lateral gradually drops off, causing the pressure loss from friction to also decline. Along the lower part of the lateral, the lower flowrates result in only slight pressure changes.

Effect of Elevation Change on Pressure

Figure 1 also shows the effect of elevation change on pressure — in this case 1% uphill and downhill slopes. Where water is flowing uphill, pressure decreases dramatically. Each one-foot increase in elevation causes a pressure loss of 0.43 psi. When the water flow is uphill, friction and elevation act together to reduce pressure along the lateral.

When the water flow is downhill, gravity helps compensate for the pressure loss caused by friction. Each one-foot decrease in elevation causes a pressure increase of 0.43 psi. Figure 1 illustrates the pressure decrease along the upper part of the lateral on a 1% downhill slope — resulting from the friction caused by the relatively high flowrates along the upper part of the lateral. In the lower part of the lateral, however, pressure increases with distance along the lateral. Here small lateral flowrate results in only slight pressure loss from friction — less than the pressure gains due to the effects of gravity — and pressure instead increases near the end of the lateral.

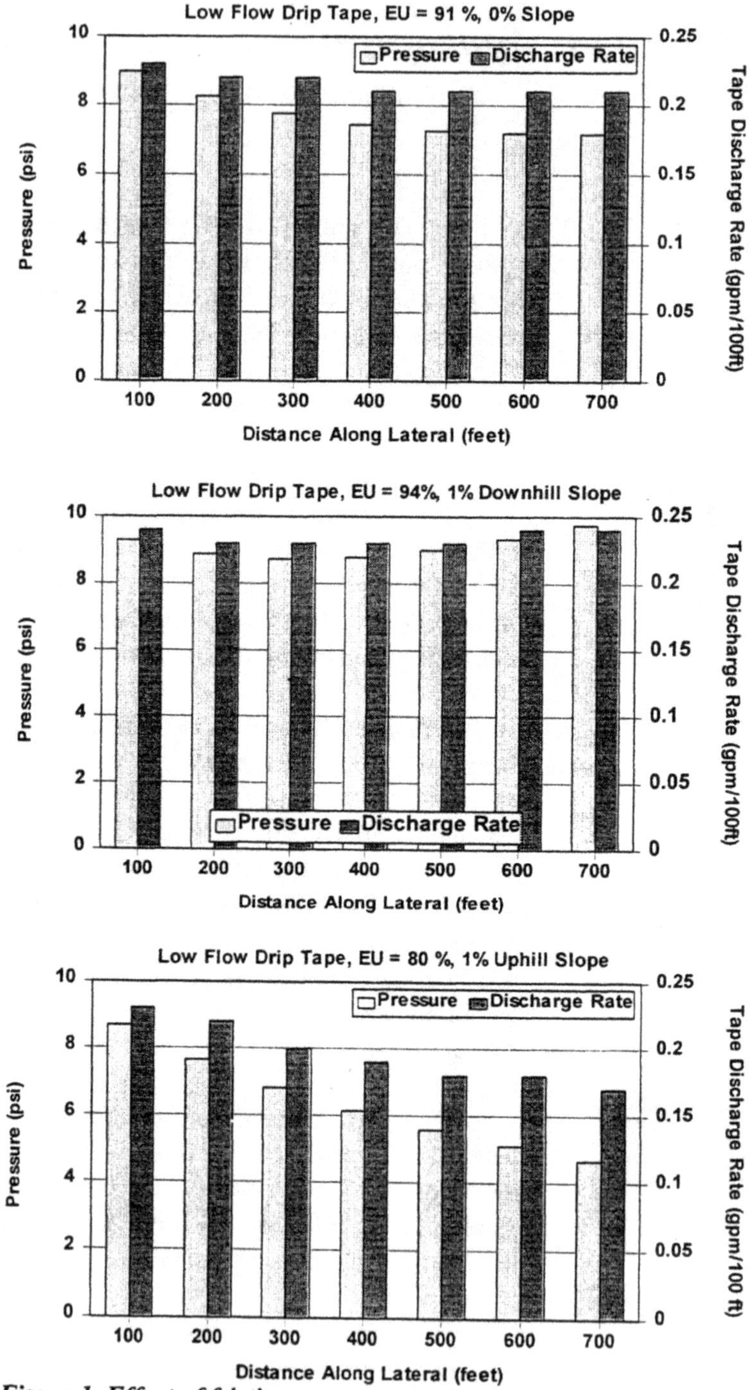

Figure 1. Effect of friction on pressure along a 0% slope, a 1% uphill slope, and a 1% downhill slope lateral.

The Relationship Between Tape Diameter and Pressure

The larger the diameter of the tape or tubing, the less the pressure loss to friction, assuming equal flowrates. Larger diameter drip tape therefore allows for longer lateral lengths without an adverse effect on emitter discharge uniformity. Doubling the tape diameter reduces the friction loss by about 29 times. Thus, friction losses are highly sensitive to tape or tubing diameter.

Flowrate and Pressure Loss

The higher the flowrate in the lateral, the greater the pressure loss from friction. Doubling the flowrate increases the friction loss by 3.6 times. The lateral flowrate depends on lateral length, emitter spacing and emitter design. The smaller the emitter spacing, the higher the flowrate for a given emitter characteristic because smaller spacings result in more emitters. The longer the lateral length, the higher the flowrate. Drip tape with "high flow" emitters will have a higher flowrate than tape with "low flow" emitters. Thus, lateral lengths should be smaller for drip tape using "high flow" emitters than for tape using "low flow" emitters.

Emitter Spacing

Emitter spacing can also affect the flowrate — and therefore the pressure — along a drip tape lateral. With a given set of emitter characteristics, the closer the emitter spacing, the higher the flowrate and the greater the pressure losses along the lateral. The closer the emitter spacing, the shorter the lateral length necessary to maintain good uniformity.

The effect of pressure on the tape or emitter discharge rate is also effected by the emitter discharge exponent. The smaller the exponent, the less effect pressure variation has on discharge rate.

Some Field Evaluations

Figures 2, 3 and 4 show tape discharge rates along the lateral length for several conditions. A high EU occurred for a 350-foot long lateral with high-flow drip tape, in Figure 2. Differences between maximum and minimum tape discharge rates were very small. In Figure 3, however, a much lower EU and much larger difference occurred along an 850-foot long lateral length with low-flow drip tape.

Figure 4 shows a lateral length of nearly 1,350 feet with an EU of 88 percent. For this case, water was flowing into both ends of the drip tape resulting in the pressure decreasing towards the middle of the lateral from both ends. This arrangement allows a longer lateral length to be used, while maintaining an acceptable EU and small differences between maximum and minimum discharge rates. A problem with this arrangement, however, in that it may not be possible to adequately flush the lateral.

Lateral Length versus EU

The effect of lateral length, emitter spacing, emitter discharge rate, and slope on the EU is shown in Figure 5. Figure 6 shows similar relationships for the ratio of the minimum discharge rate to the maximum discharge rate, or the discharge ratio. These figures, developed using software from T-Systems International, Inc, can be used to determine appropriate lateral lengths.

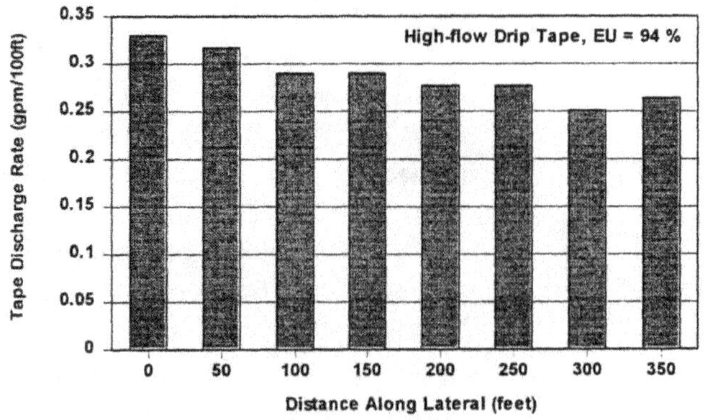

Figure 2. Tape discharge rates along a short lateral.

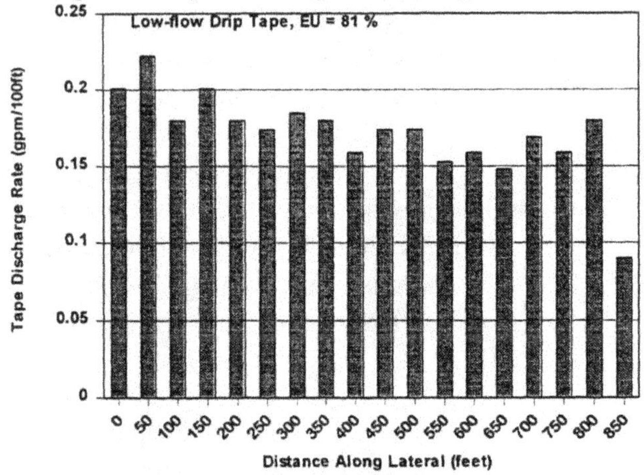

Figure 3. Tape discharge rates along a long lateral.

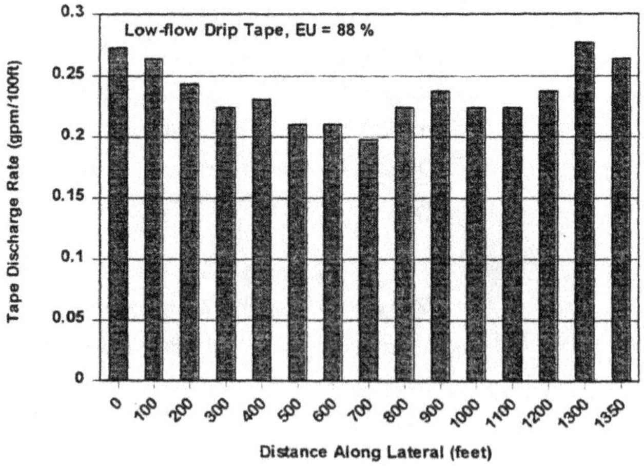

Figure 4. Tape discharge rates with water flowing into both ends of the lateral.

A design lateral emission uniformity of at least 90 percent is recommended. Thus, for a 12-inch lateral spacing, the recommended maximum lateral lengths vary from 500 feet for a high flow drip tape (0.45 gpm/100 feet) to 750 feet for a low-flow drip tape (0.22 gpm/100 feet). Smaller maximum lateral lengths occur for the 8-inch spacing and longer maximum length for the 16-inch spacing.

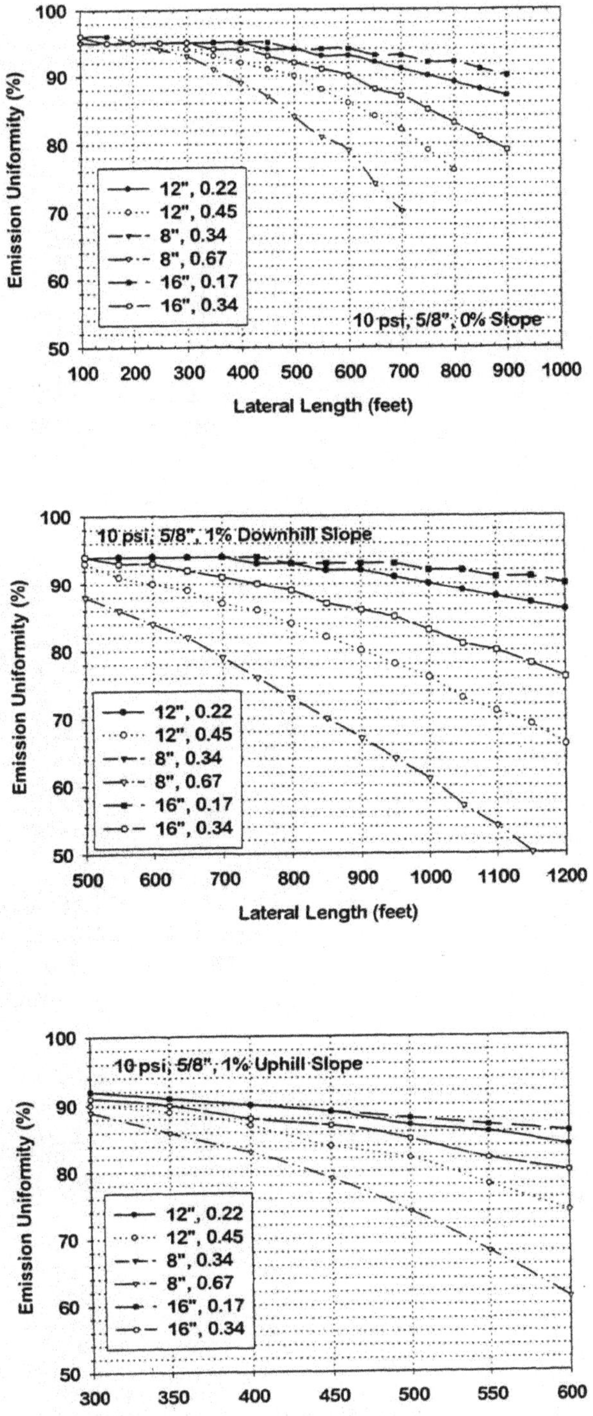

Figure 5. Emission uniformity versus lateral length for various emitter spacings, tape discharge rates, and slopes.

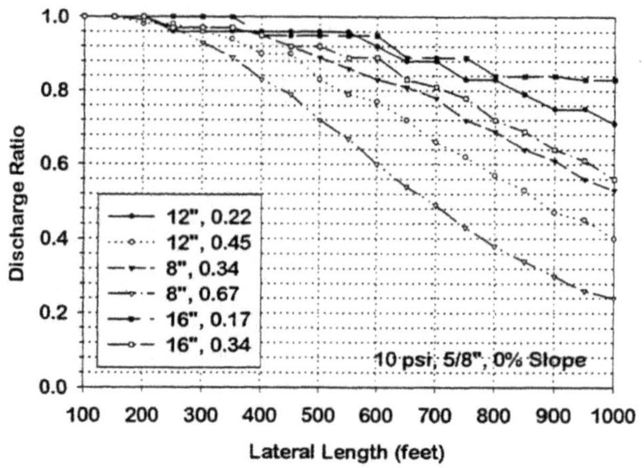

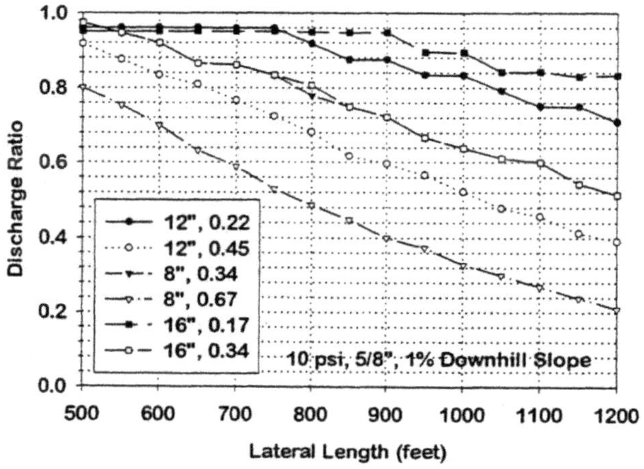

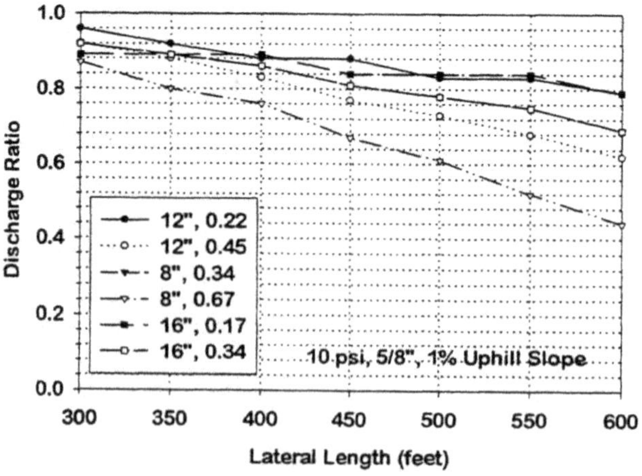

Figure 6. Discharge ratio versus lateral length for various emitter spacings, tape discharge rates, and slopes.

Tables 1 and 2 give lateral length ranges suggested for achieving a 90% uniformity, based on a survey of manufacturers. These suggested lengths take into account the effects of both the coefficient of variation and pressure changes. These tables are appropriate for drip tapes with discharge exponents between 0.4 and 0.6.

Table 1. Lateral lengths suggested for 5/8" diameter drip tape with discharge exponents between 0.4 and 0.6 to achieve 90% distribution uniformity.

Slope	Lateral length
0.2 gpm per 100 feet of length	
0	650 to 700 feet
1% (uphill)	200 to 300 feet
1% (downhill)	800 to 1000 feet
0.3 gpm per 100 feet of length	
0	500 to 650 feet
1% (uphill)	200 to 300 feet
1% (downhill)	650 feet
0.45 gpm per 100 feet of length	
0	400 to 450 feet
1% (uphill)	200 to 300 feet
1% (downhill)	550 to 650 feet

Table 2. Lateral lengths suggested for 7/8" diameter drip tape with discharge exponents between 0.4 and 0.6 to achieve 90% distribution uniformity.

Slope	Lateral length
0.2 gpm per 100 feet of length	
0	1400 feet
1% (uphill)	450 feet
1% (downhill)	1800 feet
0.45 gpm per 100 feet of length	
0	850 to 950 feet
1% (uphill)	300 to 400 feet
1% (downhill)	1150 feet

Determining the System Flowrate

By Blaine Hanson, UC Irrigation and Drainage Specialist

The design system flowrate must be sufficient to meet the beneficial water needs during the period of maximum crop evapotranspiration. This flowrate must be such that the entire field can be adequately irrigated during this peak water demand.

For other irrigation methods, the design flowrate usually is determined by first estimating the peak evapotranspiration demand and then calculating the flowrate needed to meet this demand. Factors affecting this approach include the peak water demand, the area irrigated, and the time available to irrigate the field.

Drip Irrigation Method

For drip irrigation for row crops, a different approach is recommended. A system flowrate is determined based on the lateral design and design pressure, and then the irrigation set time needed to supply the desired amount of water is calculated. A procedure for this approach is as follows:

1. Determine the design pressure and the lateral inflow rate that will occur at that pressure. The inflow rate can be calculated using either the algebraic procedure in Appendix C, manufacturer's recommendations, or a computer program that calculates pressures and emitter discharge rates along the lateral. Figures 1, 2, and 3 show lateral inflow rates for T-Tape. These relationships were developed using design software provided by T-Systems International, Inc.

2. Divide the field into a desired number of blocks. The number of blocks irrigated during each set must be sufficient to irrigate the field between irrigations.

3. Determine the number of laterals in each block, which is equal to the block width divided by the lateral spacing.

4. Determine the block flowrate by multiplying the design lateral inflow rate by the number of laterals in the block.

5. Determine the irrigate set time by the following equation:

$$T = (499 \times D \times A)/Q$$

where T is the irrigation set time in hours, A is the irrigated area of the block in acres, D is the desired depth of applied water in inches and Q is the block flowrate in gallons per minute.

6. Determine the number of blocks to be irrigated per set such that the entire field can be irrigated at the desired frequency. The system flowrate is the number of blocks irrigated at the same time multiplied by the block flowrate.

Example

Determine the design system flowrate for a 40 acre field (1,320 by 1,320 feet) irrigated daily. The maximum desired depth of applied water is 0.3 inches per day. The lateral spacing is 60 inches with an emitter spacing of 12 inches (low-flow drip tape). The tape discharge rate equals 0.22 gpm/ 100 ft.

1. The design pressure at the lateral inlet is 10 psi and the design lateral length is 650 feet (emission uniformity equals 92%). From Figure 1, the design lateral inflow rate is 1.45 gallons per minute.

2. For the first design attempt, split the field into four blocks, each block 10 acres in size. Width of each block is about 660 feet.

3. The number of laterals in each block is 660 feet divided by 5 feet (lateral spacing), which equals 132 laterals.

4. The block flowrate is 132 laterals multiplied by 1.45 gpm or 191 gpm.

5. The irrigation set time is

$$T = (449 \times 0.3 \text{ inches per day} \times 10 \text{ acres})/191 \text{ gpm} = 7 \text{ hours}$$

6. Two blocks must be irrigated during each set to irrigate the entire field each day. Total irrigation time for the field is 14 hours. The system flowrate is 2 blocks x 191 gpm per block or 382 gpm.

Considerations

Little opportunity exists to change the irrigation set time and thus the system flowrate because the drip tape discharge rate at 10 psi must be satisfied. Only by changing the design pressure can the design system flow rate be changed. However, it is recommended that the design pressure not be less than about 10 psi.

Under other irrigation methods, a design system flowrate of 6 to 10 gpm per acre is frequently recommended. Note that for this design, the unit area flowrate is 19.1 gpm per acre, considerably larger than that normally recommended. This large value again reflects the need to satisfy the lateral inflow rate at the design pressure.

The design flushing lateral inflow rate and flushing pressure should also be determined. Designs that result in flushing inflow rates and pressures much larger that the design lateral inflow rate and pressure flowrate should be avoided.

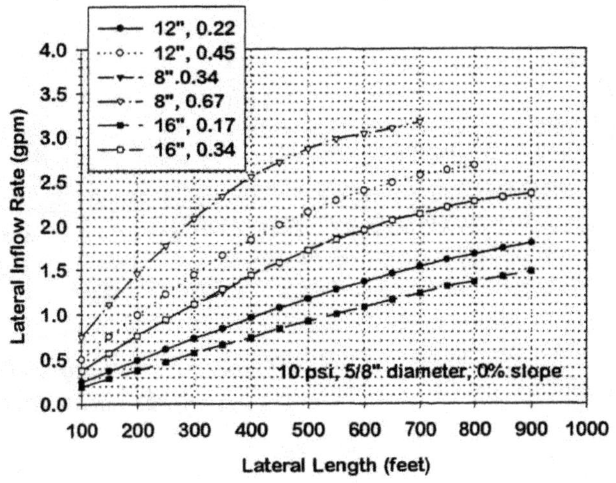

Figure 1. Lateral inflow rates for various emitter spacings and tape discharge rates for a 0% slope.

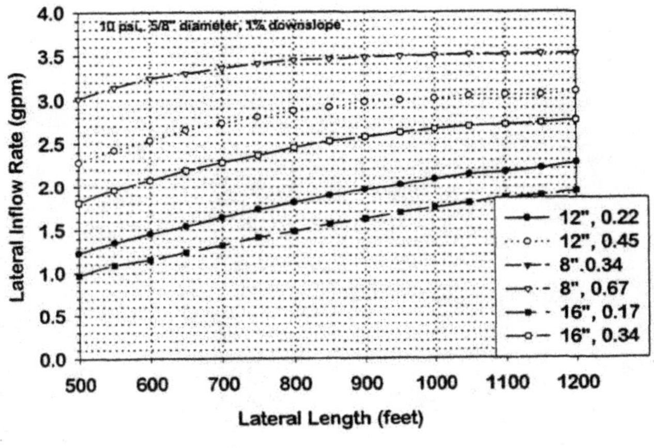

Figure 2. Lateral inflow rates for various emitter spacings and tape discharge rates for a 1% downhill slope.

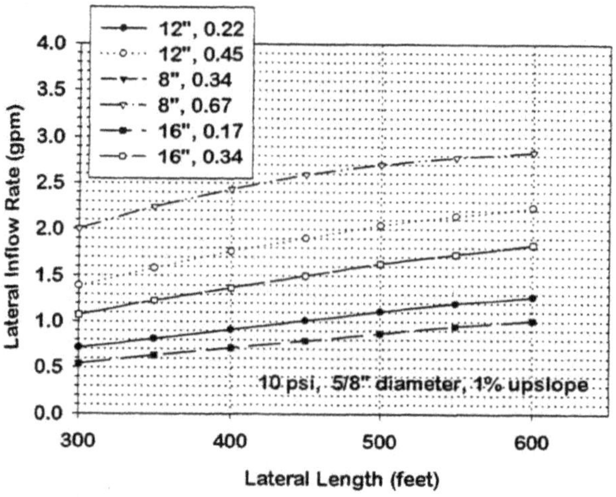

Figure 3. Lateral inflow rates for various emitter spacings and tape discharge rates for a 1% uphill slope.

Flushing

By Blaine Hanson, UC Irrigation and Drainage Specialist

Periodic flushing of drip laterals is recommended to remove sediment and other materials deposited in the drip tape, most likely along the lower part of the lateral. These deposits consist of material that passed through the filter or that precipitated out in the laterals. Flushing is accomplished by opening the ends of the laterals and allowing the resulting discharge out of the lateral to carry the deposited material with it.

The key to adequate flushing is to provide a flushing velocity of at least 1 foot per second at the end of the lateral. For a 5/8 inch diameter drip tape, a flowrate of 1 gallon per minute at the end of the lateral is required to achieve that velocity. A flowrate of 2 gallons per minute is needed for 7/8 inch diameter tape.

Figure 1 shows the lateral inflow rate needed for flushing a 5/8 inch tape. In some cases, this flowrate will be more that the design lateral inflow rate, while in other cases, the flushing flowrate may be slightly less. For example, the flushing flowrate for a 600 foot lateral using low-flow drip tape (12 inch emitter spacing) is 1.75 gallons per minute, while the design lateral inflow rate at 10 psi is 1.35 gallons per minute (see "Determining the System Flowrate"). However, for a 500 foot long lateral using high-flow drip tape (0.45 gpm/100 feet, 12-inch spacing), the flushing flowrate is 2.1 gallons per minute, while the design flowrate at 10 psi is 2.3 gallons per minute.

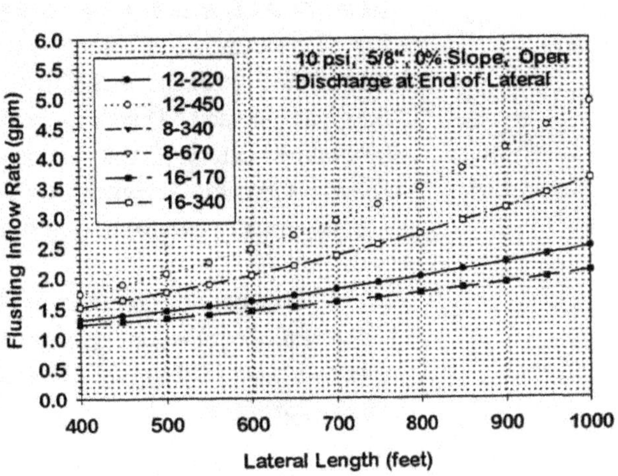

Figure 1. Flushing flowrates needed for a single lateral with an open end.

The lateral inlet pressure needed to obtain the flushing flowrate, in Figure 2, shows that 4.3 psi are needed for the 600-foot long lateral, much less that the design lateral pressure of 10 psi. This assumes that the lateral is open to the atmosphere with no back pressure such as would occur using a flushing manifold. The flushing pressure also depends on the pressure needed at the end of the lateral for flushing. A higher end pressure means a higher inlet pressure.

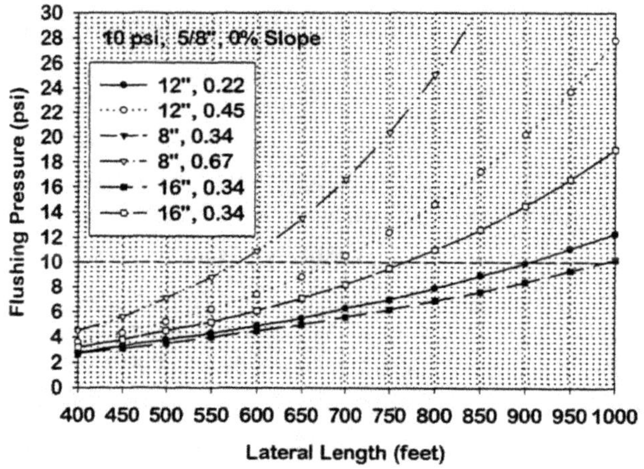

Figure 2. Lateral inlet pressure needed for flushing.

Figure 2 shows that depending of the lateral length for a given emitter type and spacing, the flushing pressure can exceed the design pressure, and in some cases can exceed the manufacturer's maximum recommended pressure. For example, a system design used by some is to supply water to both ends of the lateral by installing supply manifolds at both ends. One supply manifold is used to flush the system. Lateral lengths of about 1300 feet can be obtained using this design approach with low-flow drip tape. However, the flushing pressure required for this system can exceed 20 psi, more than the manufacturer's recommended maximum pressure.

Larger diameter manifolds, submains, and mainlines may be needed if the flushing flowrate greatly exceeds the design flowrate. However, the system flushing flowrate can also be controlled by the number of laterals flushed at a time.

Achieving the lateral flushing flowrate depends on the flushing pressure compared to the design pressure, the pump performance characteristics, and the performance characteristics of the irrigation system. Flushing flowrates that require a flushing pressure about equal to the design pressure may only be achieved by increasing the pressure in the manifold through adjusting the pressure regulator assuming that the a pump is selected that can provide the needed output.

Where the flushing pressure is much lower that the design pressure, achieving the flushing pressure depends on the number of laterals flushed at a time and the pump performance characteristics. The flushing lateral pressure can be obtained by progressively opening the ends of the laterals until the desired pressure is reached. The number of opened ends needed to reach this pressure will depend of the performance characteristics of both the pump and irrigation system.

Flushing manifolds are used at the ends of laterals to reduce the labor needed for flushing. These manifolds must be properly designed to insure that the deposited material is removed from the system. A properly designed manifold should have a water velocity of 1 foot per second. This requires a progressively increasing manifold diameter to accommodate the increasing manifold flowrate. The manifold flowrate will be the smallest at its beginning, where the flowrate equals the outflow rate of one lateral, and will be maximum at the its outlet.

Pressure-Compensating Emitters

By Blaine Hanson, UC Irrigation and Drainage Specialist

Some drip tubing emitters are designed to compensate for pressure changes in the irrigation system so long as a minimum pressure is maintained. Pressure-compensation is achieved with flexible orifices and diaphragms that change flow path dimensions in response to pressure changes. However, no drip tape with pressure-compensating emitters is manufactured.

Comparison

Figure 1 illustrates the behavior of two drip emitters, one with no pressure-compensating ability, and the other with pressure-compensating features. The discharge rate of non-compensating emitter, which has an exponent of 0.48, steadily increases as the pressure increases. The discharge rate of the pressure compensating emitter, which has an exponent of 0.0, remains constant at pressures exceeding 10 psi. If pressure falls below 10 psi, the discharge rate changes with changes in pressure.

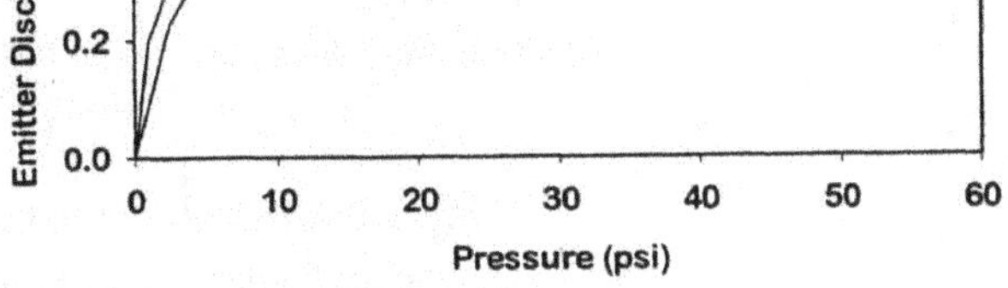

Figure 1. Comparison of a pressure-compensating drip emitter with a drip emitter having no pressure-compensating ability. *(Netafim Irrigation Company)*

Pressure-compensating emitters are particularly suited to steep or undulating terrain, which can cause significant pressure changes in the irrigation system, particularly where water flows uphill. The ability of these emitters to adjust to pressure differences can greatly even out variability in emitter discharge rates throughout the field.

Pressure-compensating tubing also makes it possible to increase the lateral length without greatly affecting the uniformity of applied water. However, as the lateral length increases, the lateral flowrates will also increase. This in turn means that more pressure is required at the lateral inlet to overcome the increased friction losses. Figure 2 shows the pressure drop along the lateral length (inlet pressure minus end pressure) required for several different emitter spacings. For relatively long lateral lengths, the pressure required to provided the lateral inflow rate may become excessive depending on the emitter spacing. In Figure 2, a pressure drop of about 30 psi will occur for lateral lengths of about 580 feet for a 12-inch emitter spacing and about 1100 feet for an 18-inch emitter spacing.

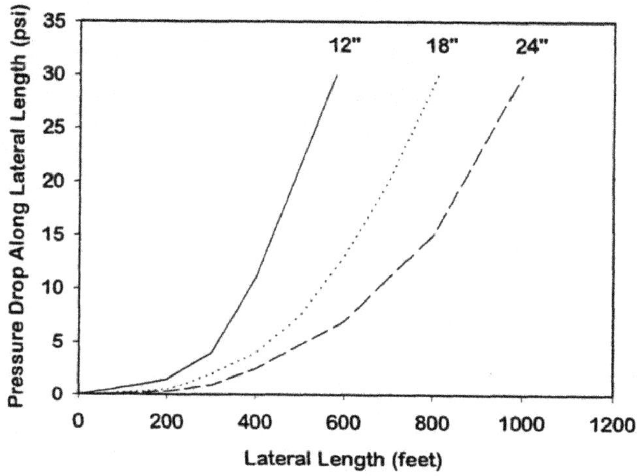

Figure 2. Pressure loss versus lateral length for a pressure-compensating emitter. *(Netafim Irrigation Company)*

Thin-walled drip tape can tolerate only very low inlet pressure — 10 to 15 psi—so that increasing the inlet pressure offers little opportunity for increasing the lateral length even if pressure-compensating drip tape were available. But drip tubing can tolerate higher pressures, making it possible for longer lateral lengths to be used with good emission uniformity resulting. At one site where drip tubing was used, the emission uniformity was nearly 95% with a lateral length of almost 1300 feet. Operating pressure was 35 psi, compared to the 10-15 psi possible with drip tape.

References

Phene, C.J., R. Yue, I-Pai Wu, J.E. Ayars, R.A. Schoeneman, and B. Meso, 1992. "Distribution uniformity of subsurface drip irrigation systems." American Society of Agricultural Engineers Paper 922569.

Schwankl, L.J., and B.R. Hanson. 1993. "Field evaluation of drip tape products." University of California Irrigation Program. University of California, Davis.

Deciding How Deep to Bury Tape

By Blaine Hanson, UC Irrigation and Drainage Specialist

Drip tape can be installed either on the soil surface, at a shallow depth of less than 12 inches, or at lower depths of more than 12 inches. Each choice offers advantages and disadvantages.

Surface Installation

Installing tape on the soil surface simplifies repairs and makes it possible to use the drip system to establish the crop. But surface installation also requires the drip tape to be removed and reinstalled with each crop and makes the system vulnerable to damage from wildlife, wind, and traffic in the field. Repeatedly installing and removing tape also requires a significant amount of labor.

Shallow Installation

Installing at a shallow depth makes the tape less susceptible to damage than surface installation and allows the drip system to be used to establish some crops. Both shallow-rooted and deep-rooted crops can be rotated with the drip system in place, but cultivation may have to be substantially modified to prevent tape damage. A sprinkler irrigation system may be needed to control soil salinity and to establish the stand.

Deep Installation

Installing at lower depths may require only minor changes in field activities to prevent tape damage, but requires a sprinkler irrigation method to establish the crop and to control salinity. Installing and maintaining the tape is more difficult at lower depths than at shallow depths and limits opportunities for growing shallow-rooted crops.

Beyond those factors, the following should be taken into account in deciding where to place drip tape:

Soil Type

Soil type. Since water moves both upward and laterally from the drip tape less readily in sandy soil than in loam soil, shallow installation may be a better choice in sandy soil.

Root Distribution

Root distribution. If the crop is shallow rooted, the drip tape should be placed at a shallow depth, preferably near the depth where the greatest number of roots are concentrated.

Seed Germination

Seed Germination. If the drip system is to be used to germinate seeds, the tape should be installed no more than 8-10 inches deep. If another irrigation system is to be used to germinate seeds, the tape can be placed deeper.

Surface Wetting

Surface Wetting. In some crops, notably melons and tomatoes, a wet soil surface can cause fruit to rot. Lowering the position of the tape can cut down on surface wetting.

Salinity

Salinity. In saline soils, salinity will be lowest close to and just below the drip tape. Placing the drip tape close to the soil surface therefore allows more of the roots access to this low salinity zone.

Crop Yield

According to one theory, the deeper the drip tape is buried the higher the crop yield because deep placement cuts down on evaporation and makes more of the applied water available to the crop. A review of studies comparing crop yield at different tape depths revealed that this theory is not necessarily valid. Some studies show that in deep-rooted crops yields are higher when tape is installed at lower depths than when tape is installed on the soil surface, but other studies show little difference in yield.

In shallow-rooted crops, the few studies available show no yield difference for the most part between surface and shallow installations, although one study on onions resulted in lower yields with a subsurface system buried 10 inches deep than with a surface drip system. It is logical to expect shallow-rooted crops to have lower yields when tape is installed at depths below the crop roots than when the tape is placed on the soil surface or at a shallow depth.

When tape is installed at lower depths, even though evaporation from the drip system may be less and the yield consequently higher, there may not be an overall water savings because a second irrigation method—sprinkler or furrow—will be needed until the crop is established. These latter irrigation methods apply water over most of the field surface, causing significant amounts of water to evaporate or to percolate below the root zone.

Deep Percolation

Deep percolation may be more severe when drip tape is placed deep in the soil than when it is placed at a shallow depth, regardless of whether a second irrigation method is used. This is particularly true if the irrigation is designed to achieve 100 percent evapotranspiration—that is, to replace all of the water used by the crop since the last irrigation. When irrigation is frequent, soil water near the emitter may exceed field capacity, and will tend to flow downward. If the tape is at a shallow depth, some of the percolating water may be taken up by the roots, but at the lower depths, the tape may be lower than most of the roots. One study conducted in Israel showed that when a drip tape was installed at 16 inches and irrigation sought to achieve 100% evapotranspiration, large amounts of water was lost to deep percolation, but that applying less water greatly cut down on deep percolation.

Usual Tape Depths

For many vegetable crops, drip tape is installed between 6 and 12 inches below the soil surface. In strawberries, a tape depth of between one and three inches is common. Some growers place tape less than six inches deep to facilitate seed germination, while others place the tape more than 10 inches deep to reduce surface wetting. Rarely is tape buried more than 12 inches below the surface, although such deep placement may be beneficial to crops such as sugar beets and potatoes.

References

Annual Report, 1986-90. Water Management Research Laboratory, U.S. Department of Agriculture (ARS), Fresno, California.

Bar-Yosef, G., B. Sagiv, and T. Markovitch. 1989. Sweet corn response to surface and subsurface trickle phosphorus fertigation. *Agronomy Journal,* Vol. 81:443-47.

Bucks, D.A., L.J. Erie, O.F. French, F.S. Nakayama, and W.D. Pew. 1981. Subsurface trickle irrigation management with multiple cropping. *Transactions of the American Society of Agricultural Engineers,* Vol. 24: 1482-89.

Constable, G.A., and A.S. Hodgson. 1990. A comparison of drip and furrow irrigated cotton on a cracking clay soil: yield and quality of four cultivars. *Irrigation Science,* Vol. 11: 149-53.

Davis, K.R., D.J. Phene, R.L. McCormack, R.B. Hutmacher, and D.W. Meeks. 1985. Trickle frequency and installation depth effects on tomatoes. In: *Proceedings of the Third International Drip/Trickle Irrigation Congress,* Fresno, California.

Plaut, Z., M. Rom, and A. Meiri. 1985. Cotton response to subsurface trickle irrigation. In: *Proceedings of the Third International Drip/Trickle Irrigation Congress,* Fresno, California.

Sammis, T.W. 1980. Comparison of sprinkler, trickle, subsurface, and furrow irrigation methods for row crops. *Agronomy Journal,* Vol. 72:701-4.

Choosing a Pump

By Blaine Hanson, UC Irrigation and Drainage Specialist

The two kinds of pumps most often used to power irrigation systems are deep-well turbines and centrifugal pumps.

Deep-well Turbines

Deep-well turbines are installed inside the well casing and consist of a pump casing, impellers (located inside the casing), and shaft, which connects the impeller to the power source. Often deep-well turbine pumps include several impeller/casing combinations (each referred to as a *stage)* to provide enough pump output.

Centrifugal Pumps

Centrifugal pumps usually consist of one impeller, a pump casing (called a *volute),* and a shaft. The shaft and the impeller rotate, creating centrifugal forces in the water inside the impeller. These forces cause water to flow to the outer edge of the impeller and into the impeller eye or center. The amount of pressure developed by the impeller depends on the impeller diameter and rpm, while the impeller flowrate is determined by the impeller width, diameter, and rpm.

Pump Performance Curves

The performance of a given pump is defined by *pump performance curves,* which describe the relationship between the total head developed by the pump at a particular capacity (the energy imparted to the water), the pump efficiency at that capacity, and the brake horsepower requirements of the pump. These curves make it possible to identify the most efficient pump for a new system by showing which pump will provide the total head and capacity the system needs. Pump performance curves can also be used to evaluate the performance of existing pumps.

Figure 1 illustrates a total head/capacity performance curve of a pump. The total head is measured in the field as the sum of the pumping lift (the distance from the soil surface down to the pumping water level) and the discharge pressure head (discharge pressure in pounds per square inch multiplied by 2.31). The figure shows that as the capacity increases, total head decreases, which is the reason pump capacity decreases as groundwater levels decline. Maximum total head occurs at shutoff (no capacity).

Total head capacity curves take a variety of shapes. Steep curves denote a slight change in pump capacity as total head changes, while flat curves denote significant changes in pump capacity with changes in total head.

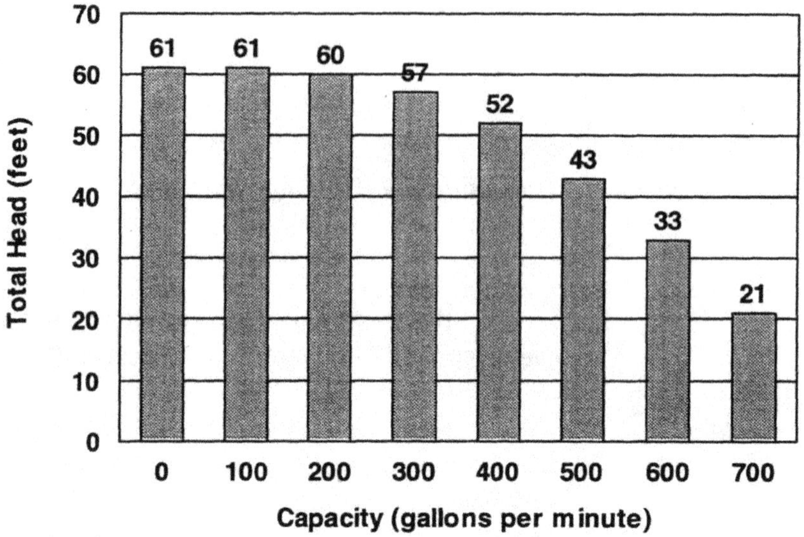

Figure 1. Total head/capacity performance curve.

Figure 2, which illustrates the pump efficiency/capacity of the same pump, shows that as capacity increases, pump efficiency also increases, until a maximum efficiency is reached. After that point efficiency decreases as capacity increases. The significance of this phenomenon is that the horse-power demand of the pump needed to provide the desired output is minimized (Note that efficiency here refers only to *pump* efficiency.) The efficiency shown in pumping plant test data is the overall efficiency—the pump and motor efficiency combined.

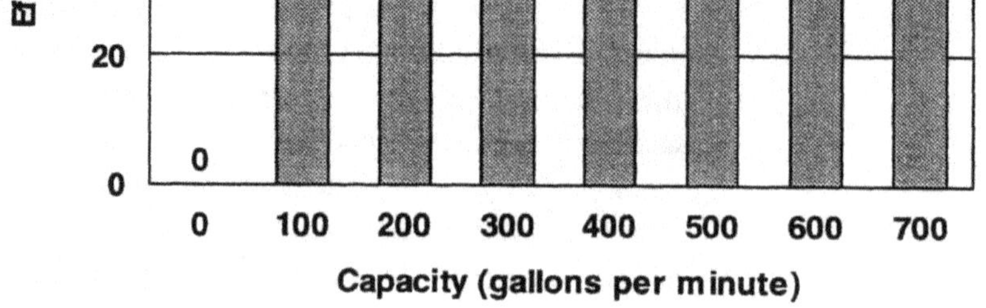

Figure 2. Pump efficiency/capacity performance curve.

The brake horsepower/capacity curve shown in Figure 3, illustrates that as capacity increases, brake horsepower also increases. In some deep-well turbine impellers, the horsepower increases to a maximum point, while in some centrifugal pump impellers, horsepower continues to increase as capacity increases.

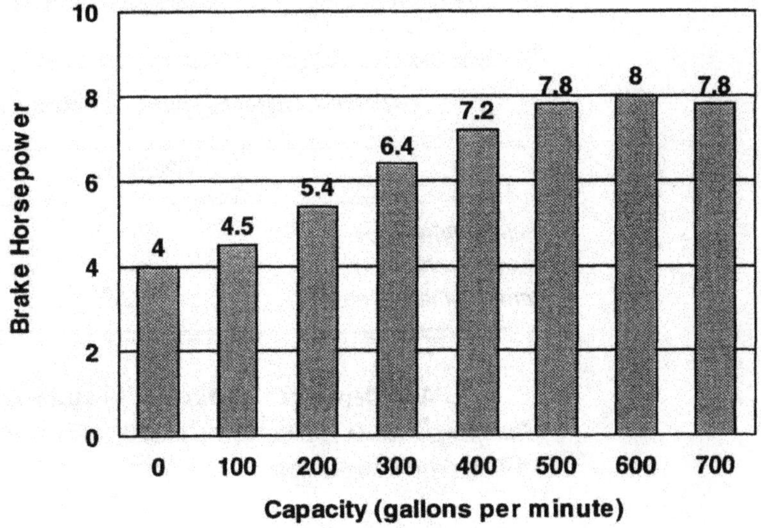

Figure 3. Brake horsepower/capacity curve.

Finding the Most Efficient Pump

The first step in identifying which pump is best suited to a particular irrigation system is measuring the pumping lift and determining the pump capacity and discharge pressure needed. The necessary capacity is determined by the crop's water requirements, the efficiency of the irrigation system, and the size of the acreage to be irrigated. The discharge pressure needed depends on the pressure requirements of the irrigation system, which in turn depends on the desired operating pressure of the drip tape, on how much pressure is lost to friction as water moves through the system, and on elevation differences in the system. The pumping lift will have to be estimated from pumping tests on the existing well or on nearby wells. The pumping lift and the discharge pressure are added together to calculate total head.

Once the total head and pump capacity have been determined, catalogs showing pump performance curves can be consulted to find a pump that will provide the desired capacity and total head at close to maximum efficiency, which will result in a minimum horsepower demand for that capacity and total head. The brake horsepower/capacity curve is used to

Example

Table 1 illustrates the importance of choosing an efficient pump. Each of the three pumps *A, B,* and *C* provides the same output (940 gallons per minute, 228 feet of total head), but the pump efficiency ranges from 69% to 84%. Brake horsepower ranges from 64 (for the 84% efficiency) to 73.5 (69% efficiency). At 1500 annual operating hours and an electric energy rate of $0.10 per kilowatt-hour, the annual energy cost ranges from $7162 (84% efficiency) to $8225 (69% efficiency).

Table 1. Effect of pump efficiency on annual energy cost.

	Pump A	*Pump B*	*Pump C*
pump efficiency	84	69	76
brake horsepower	64	73.5	69
annual energy cost ($)	7162	8225	7721

Another water management handbook in this series, *Irrigation Pumping Plants* (Publication No. 93-04), provides more detailed information about irrigation pumps.

Valves and Regulators

By Blaine Hanson, UC Irrigation and Drainage Specialist

Valves and regulators to control pressure and water flow are essential in a drip irrigation system. The following are recommended.

Check Valves

Check valves, usually installed at the pump discharge, prevent water from flowing backward into the well. These may be either of two types: swing check valves or double disc valves.

Pressure-Regulators and Pressure Regulating Valves

Pressure-regulators and pressure regulating valves dissipate upstream pressure to help keep downstream pressure constant, but do not otherwise control the water flow. These valves can be either automatic or manually operated. Figure 1 shows the performance of one type of pressure regulator.

Sometimes a gate valve is used for pressure control. These valves reduce the downstream pressure, but they do not regulate it. Any changes in the upstream pressure will change the downstream pressure where gate valves are used. A pressure regulator maintains the same downstream pressure regardless of any changes in the upstream pressure.

Pressure-Relief Valves

Pressure-relief valves, which are usually spring-loaded to open when pressure exceeds a designated level, protect against pressure surges that might damage pipelines or filter housings.

Noncontinuous Air Vents

Noncontinuous air vents allow air to escape from the pipeline while it is filling. These vents should be placed just downstream from the pump, at high points along the pipeline, or at places where air could be trapped by water while the pipeline is filling. Once the air escapes, the vent closes to prevent water from flowing out of the pipeline.

Continuous Air Vents

Continuous air vents also function to allow air to escape from the pipeline during filling, but these continue releasing air from the pipeline while the water is under pressure. These vents may be needed where air has become trapped (entrained) in the water as a result of water cascading down into the well from a point above the water level.

Vacuum-Relief Valves or Vents

Vacuum-relief valves or vents allow air to enter the pipeline after the pump is shut off and the pipeline is draining. This prevents a vacuum from forming in the pipeline.

Flow Control Devices

Flow control devices keep water flowing into the pipeline at a constant rate, despite changes in pressure. These devices usually consist of flexible orifices or membranes that change the flow passage dimensions as the pressure changes. Once the flow control device is in place the flowrate cannot be changed by varying the upstream or downstream pressure. Figure 2 shows the performance of a flow-control device. These devices normally are not used for drip irrigation systems.

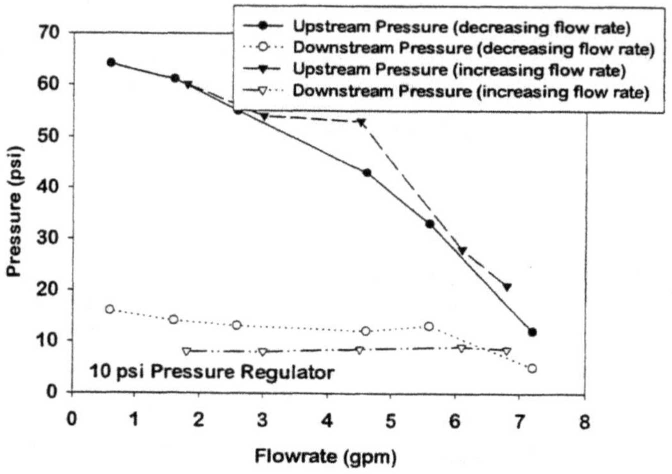

Figure 1. Performance of a pressure regulator under different flowrates and upstream pressures.

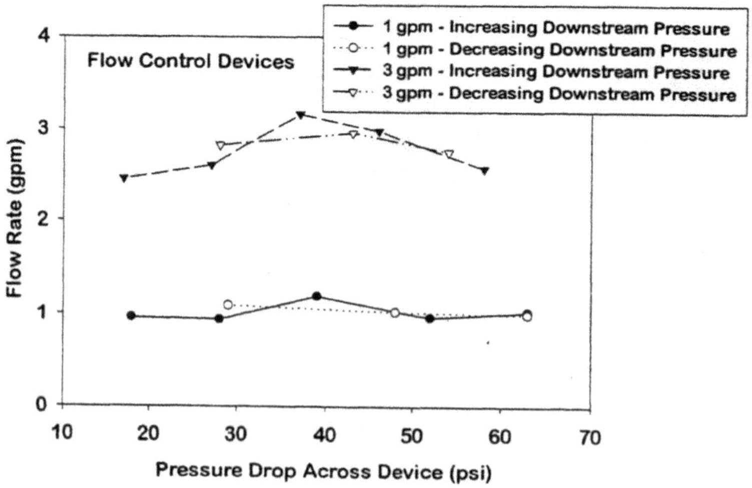

Figure 2. Performance of a flow-control device for different pressure drops.

Flowmeters: An Overview

**By Larry Schwankl, UC Irrigation Specialist
and Blaine Hanson, UC Irrigation and Drainage Specialist**

Flowmeters measure the volume of water moving through a full-flowing closed pipe and as such are one of the key components of a drip irrigation system. They are essential for managing irrigation efficiently and for monitoring the performance of the irrigation system. Managing irrigation efficiently requires: (1) knowing how much water the crop has used since the last irrigation (irrigation scheduling); and (2) operating the irrigation system to apply only the amount of water desired. A flowmeter gives the manager the information needed to apply only the amount of water required.

Monitoring the performance of a drip irrigation system makes it possible to identify changes in flowrate during the season (measured at the same pressure), which may indicate problems such as clogging of emitters or filters, leaks in the system, or problems with the pump or well.

Types of Flowmeters

Propeller Meters. The most commonly used water meters are propeller meters (Figure 1), which are installed in a straight section of the pipeline. To minimize turbulence caused by elbows, valves, and other fittings and a consequent inaccuracy in measurement, the flowmeter should be placed downstream from an unobstructed length of the pipe, eight to ten times the pipe diameter in length. For accurate measurement the pipe must flow full. Propeller meters are accurate to within two percent when new. The readout of the flowmeter is through direct-drive gears or through a cable drive (depending on manufacturer), with total and instantaneous flowrate measurements available.

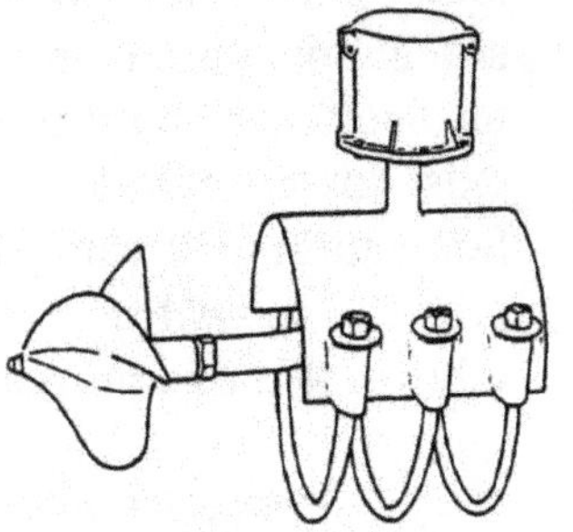

Figure 1. Flowmeter.

Propeller meters present two potential difficulties: debris in the water may entangle the propeller and affect its operation, and a small amount of pressure is lost as water flows through the meter — 2 psi or less if the meter is the proper size. Both of these difficulties are minor, however —

debris that might affect the propeller meter would also clog the emitters and therefore is usually filtered out, and pressure losses caused by the meter can be adjusted for in the irrigation system design.

Magnetic Flowmeter. The recently introduced magnetic flowmeter has the advantage of not causing an obstruction in the pipe. This feature eliminates the problem of possible entanglement from debris in the water as well as any pressure loss across the device. Magnetic flowmeters also require less maintenance than propeller meters, have long-term accuracy, and can be installed only five pipe diameters of straight pipe upstream from the meter, but have the disadvantages of a higher initial cost and the need for an external power supply.

Ultrasonic Flowmeters. Ultrasonic flowmeters measure flow velocity (and thus flowrate) by directing ultrasonic pulses diagonally across the pipe both upstream and downstream. The difference in time required for the signal to travel through the moving water is measured and converted to flow velocity. Ultrasonic flowmeters have accuracy comparable to that of propeller meters and, since they have no moving parts, require little maintenance. Because all attachments are external, these meters can be moved easily to different locations. Since the ultrasonic meters work by bouncing the ultrasonic pulses off particles in the water, irrigation waters which don't have enough suspended particles in it may not be appropriate for ultrasonic meter use. It has been found that high quality well waters may fall into this category. Ultrasonic flowmeters generally cost more than other types of meters.

Turbine Flowmeters. Turbine flowmeters operate on the principle of a rotor assembly, turning at a rate proportional to the flowrate in the pipelines. The rotor is suspended near a magnetic pickup, which records a pulse on its readout unit as each rotor blade passes. Turbine flowmeters have an accuracy comparable to that of propeller flowmeters (within 2 percent) under the correct flow conditions and require a ten-pipe diameter length of straight pipe upstream and a six-pipe diameter length downstream. The turbine flowmeter is more sensitive to non-uniform flow conditions, such as exist downstream of an elbow or constriction, than is a propeller flow meter. Some turbine flowmeters can be installed in a range of pipe diameters, which provides flexibility in system design. The applicability and cost are comparable to that of a propeller meter.

Venturi Flowmeters. Venturi flowmeters consist of a section of pipe with a restriction in a specific shape, across which pressure change is measured. The magnitude of this pressure change depends on the flowrate through the device. Venturi flowmeters offer the advantages of unobstructed water flow, no moving parts (meaning low maintenance requirements), little pressure loss across the device, and good accuracy. Venturi flowmeters cost slightly more than propeller meters and the meter readout is less convenient than that of some other flowmeters.

Propeller Flowmeters

By Blaine Hanson, UC Irrigation and Drainage Specialist

Propeller flowmeters, consisting of a propeller linked by a cable or shafts and gears to a flow indicator and inserted into a pipeline, are frequently used to measure flowrates in pipelines. The indicators can report either the flowrate or the total flow volume, or both.

Propeller flowmeters can be installed in several different ways: by being inserted into a short section of pipe, which is then either coupled(Figure 1a); bolted (Figure 1b) into the pipeline; clamped, strapped, or welded onto the pipeline as a saddle-type meter (Figure 1c); or inserted into the pipe discharge (Figure 1d) to measure flow from gates installed in canals and ditches.

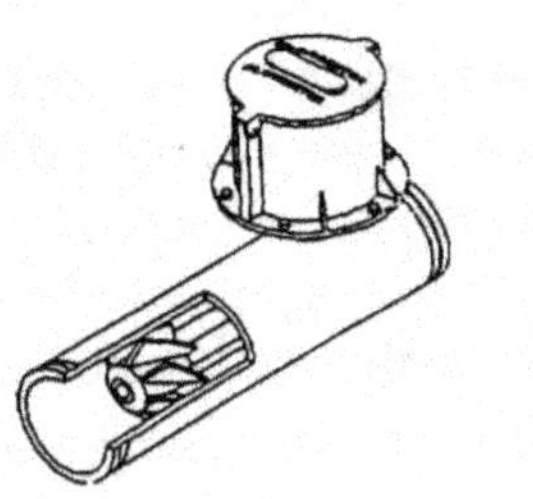

Figure 1a. Grooved propeller flowmeter

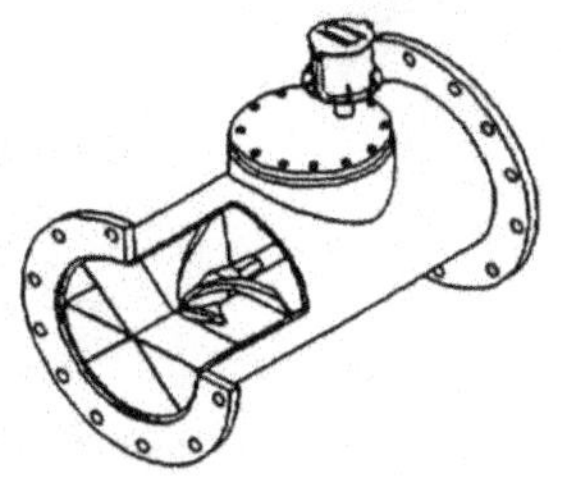

Figure 1b. Flanged propeller flowmeter with straightening vanes

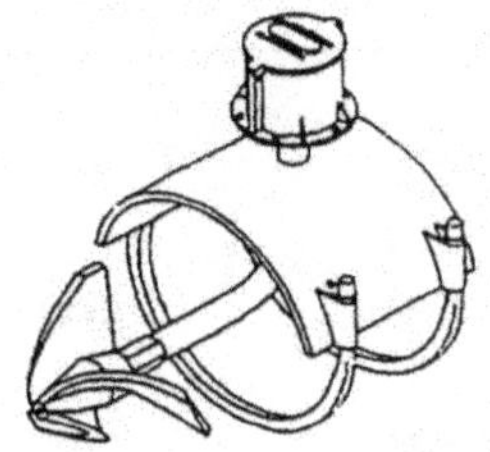

Figure 1c. Saddle-type propeller flowmeter

Figure 1d. Open-flow propeller flowmeter

(Illustrations courtesy of Ketema/McCrometer Division Hemet, CA 92545.)

Propeller meters can also be installed as in-line meters in a short section of portable pipe (Figure 1a). Couplings are welded onto each end of the flowmeter pipe section to connect sections of the portable pipe. One flowmeter of a given size can be used for several sizes of portable pipe if a sufficiently long straight section of pipe of the same diameter as the flowmeter section is installed immediately upstream from the flowmeter.

Since clamp-on and strap-on saddle meters can be moved from location to location, one flowmeter can be used to measure flowrates for several pumping plants. The same pipe diameter must be used at each site, however. Dummy saddles can be installed at locations other than the one at which the meter is installed.

Propeller meters must be matched to the correct pipe size — since the gear mechanism connecting the propeller to the indicator is based on the pipe diameter — and to the desired flowrate. Table 1 gives maximum and minimum flowrates for various pipe diameters.

Most flowrate indicators report in gallons per minute or in cubic feet per second, while total flow indicators ("totalizers") report in gallons, acre-feet, or cubic feet. Some indicators report in metric units.

The flowmeter should be installed at a location of minimal water turbulence, since too much turbulence will cause the flowrate indicator to oscillate wildly, preventing reliable measurement.

Manufacturers often recommend that a section of straight pipe, eight to ten pipe diameters long, be placed immediately upstream from the flowmeter, but field experience has shown that a longer pipe length may be required in locations where jetting may occur because of a partially closed valve. One manufacturer maintains, however, that reliable measurements can be made with even relatively short sections of pipe if straightening vanes are used.

Table 1. Maximum and minimum flowrates of various pipe diameters.
(Courtesy of McCrometer Flowmeters)

Meter and nominal pipe size	4	6	8	10	12	14
Maximum flowrate	600	1200	1500	1800	2500	3000
Minimum flowrate	50	90	108	125	150	250

Swirling

A centrifugal sand separator or a series of elbows in the pipeline may cause a swirl or rotation to develop in the flowing water. The swirls may still be present even a hundred pipe diameters downstream. The remedy is to place straightening vanes in the pipeline just before the flowmeter. One manufacturer recommends that a six-vane straightener be used.

A relatively stable propeller meter rate indicator means that turbulence is minimal, but wide variation in indicator readings signals turbulence in the pipeline. If the rate indicator shows erratic, violent behavior, air or gas may be present in the water.

Attaining Full Flow

Propeller flowmeters will operate properly only if the pipe is flowing full. If the pipeline is only partially full, the flowrate measurement will not be accurate. In pressurized irrigation systems, flow will usually be full at the pump discharge, but in pumps with an open discharge, as into an irrigation ditch, pipe flow may not be full. The problem can be remedied by creating a slight rise in the discharge pipe; by installing a "gooseneck" at the pipe discharge; by installing an elbow (discharge end pointing upward at the pipe discharge); or by installing a valve downstream from the meter.

Under ideal conditions, a propeller flowmeter operated within its recommended range can be accurate to within ± 2 percent, but if the flowrate is too slow, accuracy will be less.

Accuracy

Figure 2 shows that accuracy will remain within ± 2 percent as long as the flowrate exceeds about 200 gpm for a ten-inch meter, assuming there is no excessive turbulence immediately upstream from the meter.

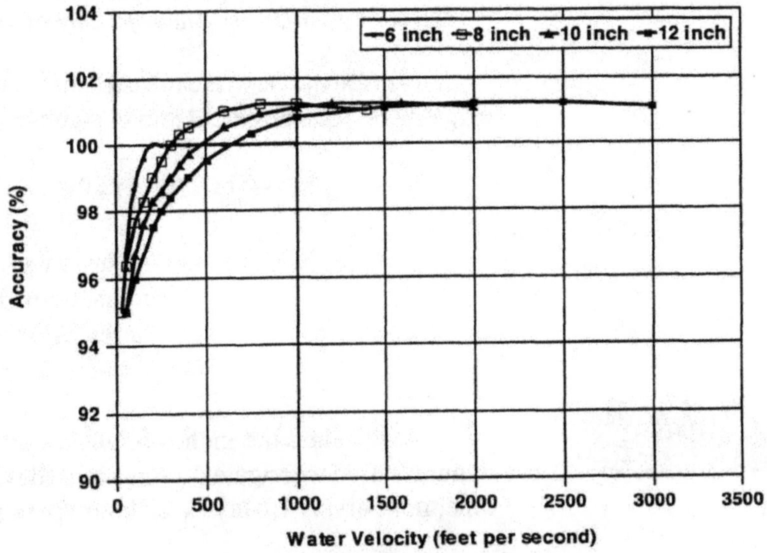

Figure 2. Propeller flowmeter accuracy.

(Figure developed from data supplied by Water Specialties Corporation, Porterville, CA 93257.)

Inserting a propeller into the water flow can cause friction, resulting in pressure or head losses in the pipeline. The amount of pressure lost depends on the velocity or flowrate and on the pipe diameter. The higher the flowrate, the more pressure lost because of the flowmeter, but the larger the pipe diameter, the less pressure lost.

As Figure 3 illustrates, these pressure losses are generally small. With a ten-inch flowmeter, the pressure loss is less than 0.1 psi for flowrates less than 2000 gpm.

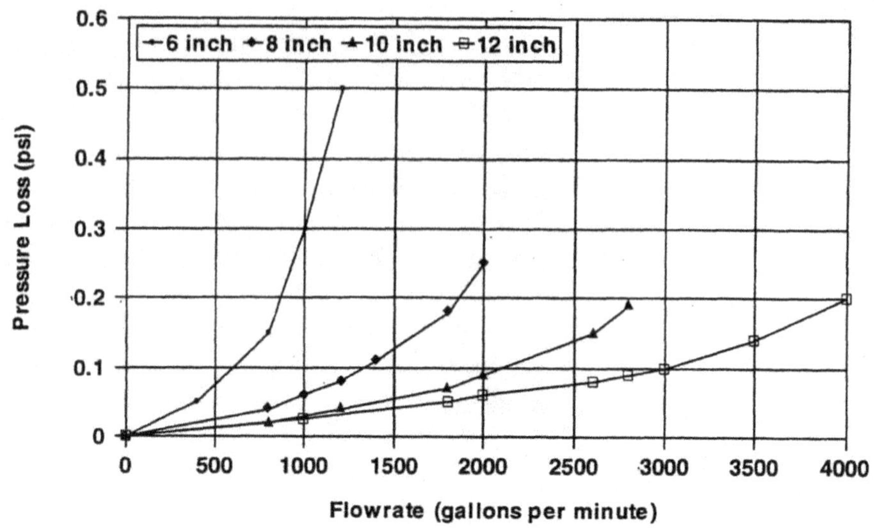

Figure 3. Pressure loss caused by propeller flowmeters.
(Figure developed from data supplied by Water Specialties Co, Porterville, CA 93257.)

From the flowrate measurement, the amount of water applied during an irrigation set can be calculated in inches using the following equation:

$$D = (Q \times T) \div (449 \times A) \tag{1}$$

where D = inches of applied water,
T = actual hours required to irrigate the field
A = acres irrigated
Q = flowrate in gallons per minute

Example

Calculate the inches of water applied if a flowrate of 300 gallons per minute is used to irrigate 40 acres in 10 days at 22 hours per day. Total irrigation time, in hours, is 10 days x 22 hours per day = 220 hours.

$$D = (300 \text{ gpm} \times 220 \text{ hours} \div 449 \times 40 \text{ acres}) = 3.7 \text{ inches}$$

References

Ketema/McCrometer Division. *McCrometer propeller flowmeters: manual for installation, operation, and maintenance.*

______________. *Bulletin G1200. Basic specifications for McCrometer propeller flowmeters.*

______________. *Bulletin MR-300. McCrometer flowmeters.*

Noffke, Milvern H. 1991. "Achieving accurate irrigation water measurements with propeller meters." ASCE Specialty Conference, *Planning Now for Irrigation and Drainage in the 21st Century,* July 18-21, 1988, Lincoln, Nebraska.

Filtration Equipment: An Overview

By Larry Schwankl, UC Irrigation Specialist
Blaine Hanson, UC Irrigation and Drainage Specialist
and Terry Prichard, UC Water Management Specialist

Since drip emitters have small passageways that can become clogged by inorganic materials (sands, silts, clays), and the chemical precipitates (lime and iron) or by organic materials (algae, bacteria, and the slimes), irrigation water must be filtered if it is to be used in a drip system. Chemical precipitates are usually associated with groundwater sources, while organic materials are most common in surface waters.

Sand, silt, plant material, algae, fish, and other materials suspended in the water must be removed before the water enters the drip line. Removal of these materials is accomplished by filtering the water. Any particles larger than about one-tenth the size of the emitter orifice or flow passage should be filtered out to prevent particles from jamming.

Types of Filters

Types of filtration equipment for removing suspended materials are as follows:

- Suction screens installed on intake of centrifugal pumps.
- Centrifugal sand separators.
- Sand-media filters.
- Gravity screen filters.
- Pressurized screen filters.
- Disc filters.
- Cartridge filters (used on small irrigation systems).

Guidelines

The degree of filtration is expressed as the mesh size or equivalent mesh size. The mesh size is the number of openings per inch. A 200 mesh filter has 200 openings per inch. A common recommendation for filtration is a 200 mesh filter.

Two sets of filters are recommended. The primary filters remove suspended material from the water. A secondary filter installed downstream from the primary filter is recommended to protect the irrigation systems if the primary filter should fail.

Table 1 lists the soil particles and their range of equivalent mesh sizes. Note that a 200 mesh filter will not remove particles of fine sand, silt, and clay. These particles will pass through the filter and be deposited in the drip lines. Flushing will be required to prevent clogging from sedimentation in the drip lines.

Which Filter is the Best?

The best filter depends on the type of suspended material in the irrigation water. Generally, sand-media filters are used to remove algae and other organic material. These filters are effective in removing these materials without the need for frequent backflushing. Screen and disc filters can remove organic material, but they can clog rapidly, thus requiring frequent backflushing. In the case of screen filters, a cake of organic material can form on the surface of the screen. Some studies have shown that during the irrigation, particles in this cake can squeeze through the openings in the screen resulting in a higher concentration of suspended organic solids in the filtered water than in the inflow water.

Several studies comparing filters where organic material occurred in the irrigation water showed the following:

1. The rate at which the pressure loss increased across the filter was less for sand-media filters than screen and disc filters. Several studies showed, however, that for small particles of organic material, differences in the rate were small for the sand-media and disc filters.

2. The rate of pressure loss across disc filters was less compared with screen filters. However, one study showed that where large organic particles (1 mm on size) occurred, the pressure loss across the disc filter increased faster compared with screen particles.

3. The larger the mesh size, the faster the increase in pressure loss across the filters.

References

Hanson, B.R. and L.J. Schwankl. 1996. *Comparison of sand-media and disk filters for micro-irrigation.* Final report submitted to the UC Committee on the Relationship of Energy to Aquaculture.

Adin, A. and E. Menachem. 1989. Particle filtration for wastewater irrigation. *Journal of Irrigation and Drainage Engineering,* American Society of Civil Engineering, Vol 115(3): 474-487.

Adin, A. and G. Alan. 1986. Mechanisms and process parameters of filter screens. *Journal of the American Society of Civil Engineers,* Vol 112(4):293-304.

Racina, I., E. Paz, G. Sagi, A. Schischa, A. Marcu, Z. Yechiely, Z. Sofer and Y. Leu. 1995. Performance evaluation of filters and emitters with secondary effluent. *Proceedings of the Fifth International Micro-irrigation Congress,* Orlando, Florida, April 2-6, 1995.

Table 1. Particle size classification by mesh size.

Very Coarse Sand	18 to 10 mesh
Coarse Sand	35 to 18 mesh
Medium Sand	60 to 35 mesh
Fine Sand	160 to 60 mesh
Very Fine Sand	270 to 160 mesh
Silt	400 to 270 mesh
Clay	Smaller than 400

Table 2. Filtration guidelines.

Flowrate	Concentrations Organic	Inorganic	Filtration*
less than 50 gpm	L	L	A
	L	M	C+A
	L	H	C + A
	M	L	D or A
	M	M	C + D, or C + A
	M	H	C + D, or C + A
	H	L	D or A
	H	M	C + D or C + A
	H	H	C + D or C + A
50 - 200 gpm	L	L	A
	L	M	C + A
	L	H	C + A
	M	L	B or E
	M	M	C + B or C + D
	M	H	C + B, C + D or C + E
	H	L	B, D or E
	H	M	C + D, C + D or C + E
	H	H	C + B, C + D or C + E
more than 200 gpm	L	L	A
	L	M	C + A
	L	H	C + F or E, C + A or C + E
	M	L	B or E
	M	M	C + B or C + E
	M	H	C + B or C + E
	H	L	B or E
	H	M	C + B or C + E
	H	H	C + B or C + E

L = less than 5 ppm
M = 5 - 50 ppm
H = more than 50 ppm

A = screen/disk filter
B = suction screen filter
C = centrifugal separator
D = gravity filter
E = sand media filter

* Letter sequence indicates the sequence of the filters: C + E means a centrifugal sand separator followed by a sand-media filter.

Centrifugal Sand Separators

By Blaine Hanson, UC Irrigation and Drainage Specialist

Centrifugal sand separators can be used to remove larger particles of sand. Water entering the sand separator rotates around the separator. This rotation develops centrifugal forces that cause sand particles to move to the outer edge of the separator. These particles then settle out of the water into a collection chamber at the separator's bottom. Clean water flows out of the separator at the vortex of the rotation, along the axis of the separator. Figure 1 shows two configurations of sand separators.

Sand separators are effective in removing up to nearly 98 percent of particle sizes larger than an equivalent 200 mesh. Organic material will not be removed by these devices.

Sand separators must be sized according to flowrate for proper operation. A properly selected sand separator will have a pressure drop of 5 to 11 psi. Manufacturer's charts on flowrate verses pressure drop such as shown in Figure 2 are used to size centrifugal separators. A flowrate that is too small will not be effective in removing sand; a flowrate that is too large will cause excessive pressure losses.

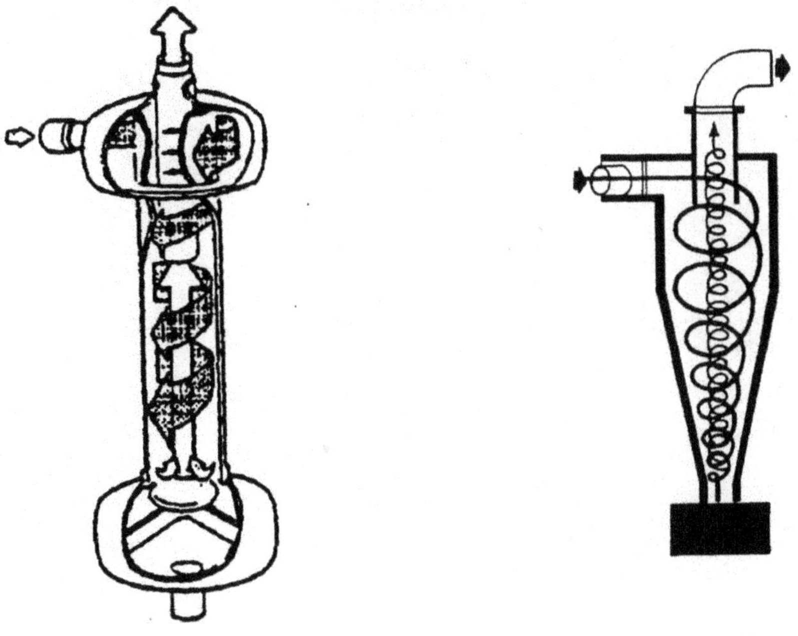

Figure 1. Centrifugal sand separators.
(Claude Laval Corporation and Netafim Irrigation Company)

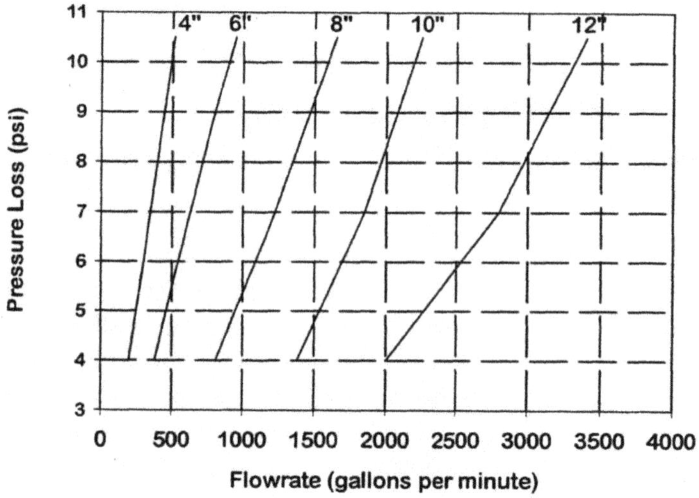

Figure 2. Pressure loss versus flowrate.
(Clemons Sales Corporation)

Varying flowrates can result in an ineffective operation of sand separators. Varying-sized fields with the same pump, declining ground water levels, or irregularly shaped field for which the number and length of laterals change during the field irrigation can cause the flowrate to vary.

Centrifugal sand separators are normally installed on the discharge side of the pump. They can also be installed in a well on the intake side of a deep-well turbine.

Sand-Media Filters

By Blaine Hanson, UC Irrigation and Drainage Specialist, and Larry Schwankl, UC Irrigation Specialist

Sand media filters are tanks filled with silica sand or crushed granite. Contaminants are filtered from the water as the water flows through the filtering media, shown in Figure 1. Tanks are usually made of carbon steel or stainless steel. One company now offers tanks made with a composite material that greatly reduces the weight of the tank. At least two tanks are required because the filtered water from one tank is used to backflush the other tank.

Water flowing through the filter is collected at the bottom of the tank by an underdrain system. One type of underdrain is an epoxy-aggregate composite. Another type consists of slotted or perforated screens at the bottom of the tank covered with gravel.

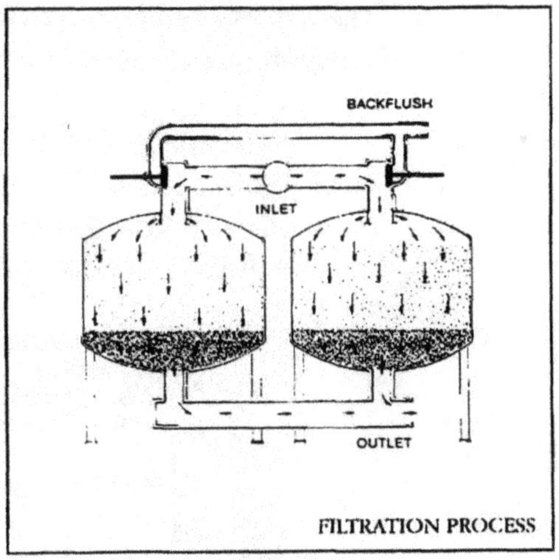

Figure 1. Filtering mode.

The type and size of sand-media affects the filtering capacity of these filters. Sand sizes are designated by the sand number with a number 8 sand having the largest sand size and a number 30 sand having the smallest sand size. Table 1 lists the sand numbers and size and the equivalent mesh sizes.

Table 1. Sand media size and screen mesh designation.

Sand No.	Effective Sand Size (inches)	Equivalent Mesh Size
8	0.059	70
11	0.031	140
16	0.026	170
20	0.018	230
30	0.011	400

The filtering ability also depends on the type of media. Figure 2 shows that crushed silica filters better than does Monterey sand because the angular particles of the crushed material fit closer together than do the sand particles. For a given sand number, the minimum particle size retained is much smaller for the crushed silica than for the Monterey sand.

Design Flowrate

The recommended design filter flowrate depends on the amount of suspended materials in the water. For dirty water (100 ppm or more of suspended material), an acceptable filter flowrate is about 15 gallons per minute per square foot of filter area. Recommended design flowrates of an "average water" are 20 to 25 gallons per minute per square foot of filter area. The recommended design flowrate for clean water (less than 10 ppm of suspended solids) is 25 to 30 gallons per minute per square foot.

Excessive filter flowrates can reduce the filtering ability of sand-media filters. Larger flowrates result in larger particles passing through the sand-media. Figure 2 shows the minimum particle size retained for different sand-media over a range of filter flowrates.

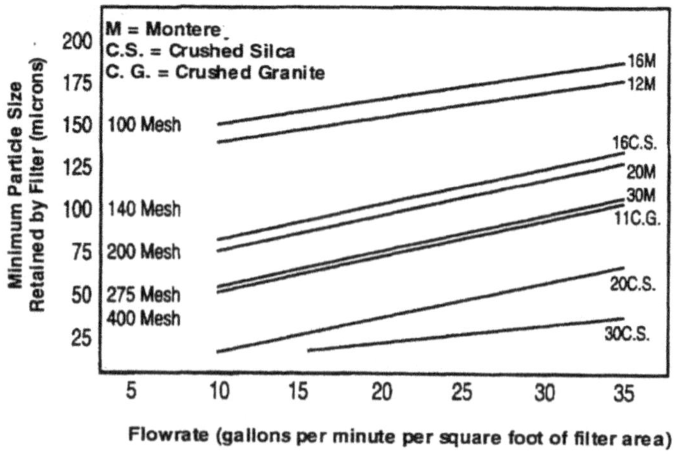

Figure 2. Minimum particle size retained for various sand sizes and flowrates. *(Everfilt Corp.)*

The design pressure drop in the filter is 3 to 5 psi. Filters should be backflushed when the pressure drop reaches about 10 psi.

Backflushing

Backflushing of filters should occur when the pressure drop becomes about 10 psi. Backflushing involves reversing the flow through the filter using clean water from the other filters to remove the contaminants, shown in Figure 3. The backflushing flowrate should be sufficient to fluidize or suspend the sand-media. This allows the contaminant to be carried out by the backflushing water. Recommended backflushing flowrates are 10 to 15 gallons per minute per square foot of filter area for Number 30 and 20 sands and 20 to 25 gallons per minute per square foot for Number 16 and 11 sands. The actual backflushing flowrate should be determined from visual observation. The flowrate should be sufficient such that a very small amount of sand-media is carried out during backflushing. A pressure-sustaining valve may be needed downstream of the filters to ensure sufficient backflushing flowrates.

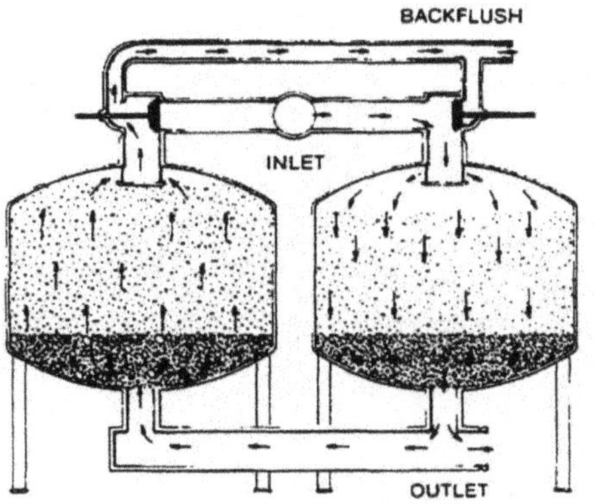

BACKWASHING PROCESS

Figure 3. Backflushing of sand-media filters.

During backflushing, the flowrate to the irrigation system can decrease. This can only be avoided by oversizing the filters and selecting a pump such that during backflushing the flowrate through the remaining filters equals the desired irrigation system flowrate and the backflushing flowrate.

Removing fine sand may be difficult particularly for two-filter systems. Because the specific gravity of the fine sand may be similar to that of the sand-media, the fine sand may not be removed unless the backflushing flowrate is increased to a flowrate higher than normally recommended. Compacted layers of fine sand in the other tanks also may prevent a sufficient flowrate for backflushing particularly in two-tank systems.

Chlorination

Periodic chlorination of sand-media filters may be required. Biological matting can occur in the filters which can severely degrade their performance. These mats also may prevent adequate backflushing flowrates. Also, during the offseason, organic material may grow in the filters. Chlorination of the filters using contact times of several hours at chlorine concentration of 10 to 20 ppm has successfully restored filter performance.

Screen and Disc Filters

By Blaine Hanson, UC Irrigation and Drainage Specialist

Screen Filters

Screen filters normally are used to filter inorganic suspended materials such as sand and other large particles. These filters will also remove algae and other organic matter but tend to clog rapidly. Under conditions of organic loads, screen filter with automatic backflushing are recommended.

Types of screens include perforated screens, wedgewire screens, and weavewire screens (Figure 1). Perforated screens are coarse-meshed, used for pretreatment of the water. Fine-meshed screens are of the weavewire type, used for primary and secondary filtration. Available mesh sizes range between 20 and 450 mesh.

Figure 1. Types of screens.
(Amiad Water System Technologies)

Maximum recommended flowrates generally range between 150 gallons per minute per square foot of filter and about 200 gpm per square foot depending on the manufacturer. The pressure loss through a screen of a clean filter depends of the flowrate as shown in Figure 2 for one filter type. Recommended design pressure loss is 3 to 5 psi.

Recent research revealed the following about the screen filter. (Zeier, K. and D.J. Hills. 1987. Trickle irrigation screen filter performance as affected by sand size and concentration. *Transactions of the American Society of Agricultural Engineers,* Vol. 30(3): 735-739.)

1. The primary factor affecting clogging of screen filters by sand is the sand size. A secondary factor is the concentration of the sand in the irrigation water.

2. Fine sands cause a faster pressure drop across a screen filter than coarse sands at similar concentrations. This suggests that more filter area may be needed where fine sands occur in the irrigation water.

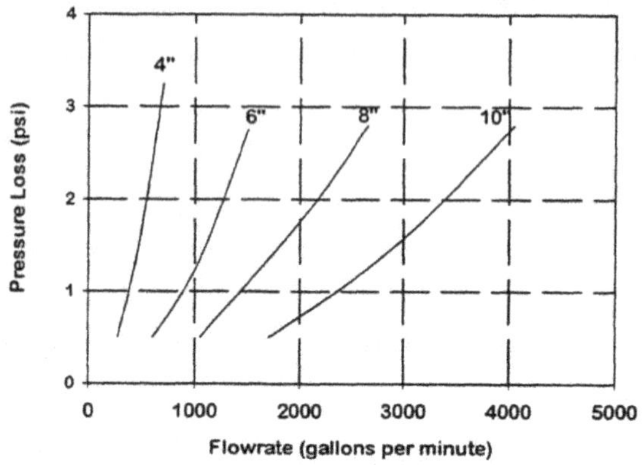

Figure 2. Pressure loss across a screen filter.
(Thompson Filters)

A variety of filter configurations are available. A brief description of some of these screen filters follows:

1. Water flows into the inside of a tubular screen and passes through the screen to the outside of the screen as shown in Figure 3. Backflushing is accomplished by opening a backflush port opposite the inflow port. Water flowing through the screen cleans the screen.

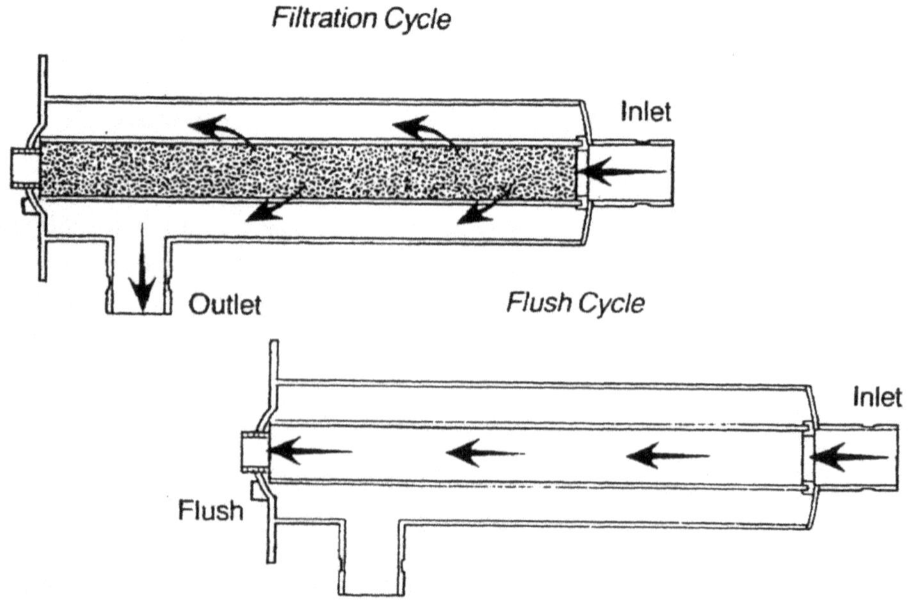

Figure 3. Tubular or blow-down screen filter.
(Flow-Guard Filtration Products)

2. Figure 4 shows a configuration in which water flowing into the screen swirls around the screen. This turbulence is designed to help keep the screen clean. Debris removed by the turbulence collects at the bottom of the screen. A continuous discharge at the bottom of the screen is suggested. A modification to this configuration is where swirling water enters the outside of the screen.

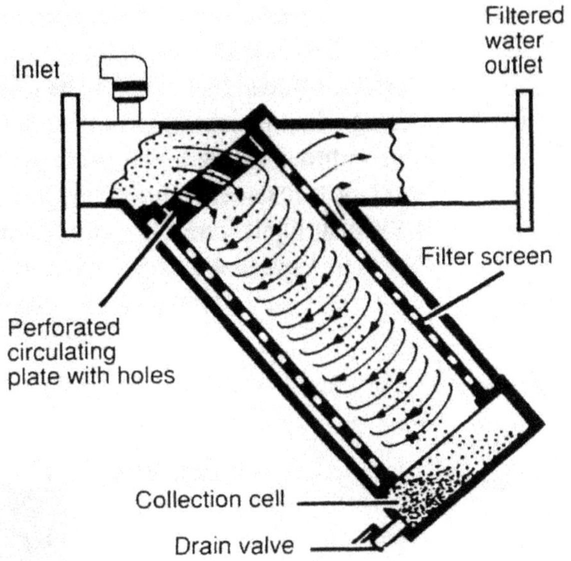

Figure 4. Screen filter with spiraling water.
(Netafim Irrigation, Inc)

3. A cone-shape screen is used for the configuration in Figure 5. Water enters through the bottom of the filter where the water velocity decreases allowing some suspended material to settle out in the bottom of the filter.

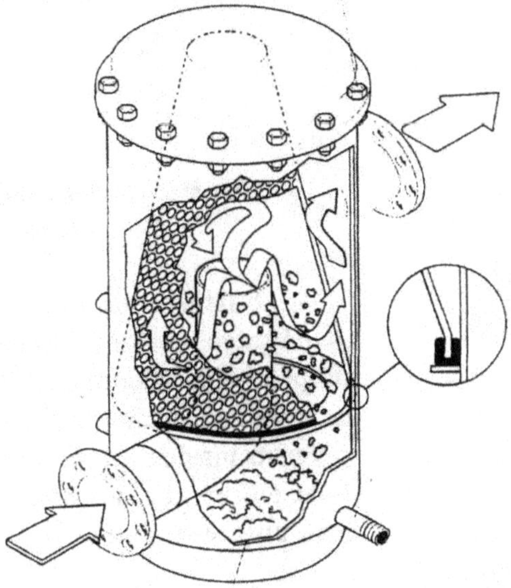

Figure 5. Conical-shaped screen filter.
(Thompson Filter, Miller-Leaman Inc.)

Several companies manufacture screen filters with automatic backflushing capabilities. One configuration consists of a bank of two or more screen filters similar to the filter in Figure 3. During the backflushing, the flow is reversed using clean water from the other filters. The backflush water is discharged through the backflush port.

Several companies make screens similar to that shown in Figure 6. Water first flows through a coarse screen for pretreatment and then passes through a fine screen. Inside the fine screen are nozzles that when their flow passage are opened to the atmosphere, clean water flows back through the screen into the nozzle. The potentially large pressure drop across the screen at the nozzle causes a jet of water to flow through the screen, thus backflushing it. One company estimates that the volume of backflush water is less than 1 percent of the total flow.

Figure 6. Screen filter with automatic backflushing capability. (Amiad Water System Technologies)

Suction Screens

Suction screen filters are coarse screens used for prescreening and are placed at the inlet of the pump intake pipe when surface water is being pumped. The filters should be used where concentrations of organic materials are moderate. Mesh sizes range from 10 to 30. These systems use rotating water jets inside the screen for self-cleaning. Pressure for the jets is supplied by the discharge side of the pump. Flowing water is needed to remove trash jetted off the screen.

Gravity Screens

Several companies market gravity-flow screens. One design, in Figure 7, uses rotating jets to keep the screen free of debris. Another design uses a screen installed with a steep slope, shown in Figure 8, to keep the screen from clogging.

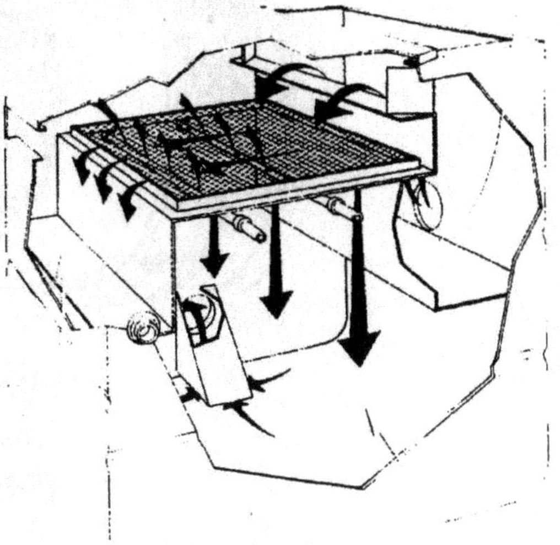

Figure 7. ***Gravity screen filter with jets.***
(Fresno Valves and Casting, Inc.)

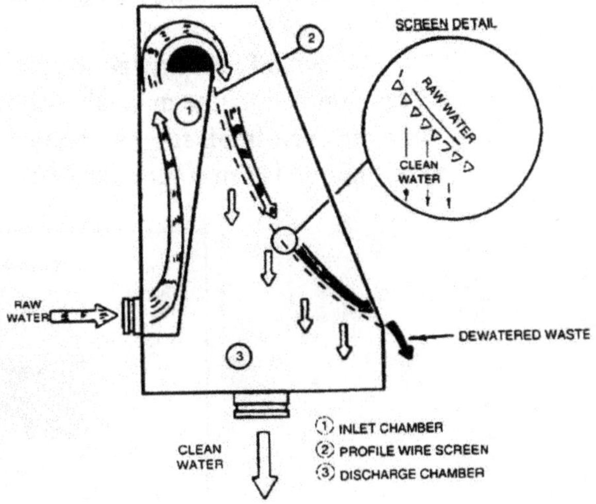

Figure 8. ***Gravity screen filter.***
(Everfilt Corp.)

Disc Filters

Disc filters consist of a series of plastic rings containing microgrooves, shown in Figure 9. Water flows from the outside of the rings through the microgrooves to the inside of the rings. Equivalent mesh sizes range between 40 and 600 mesh.

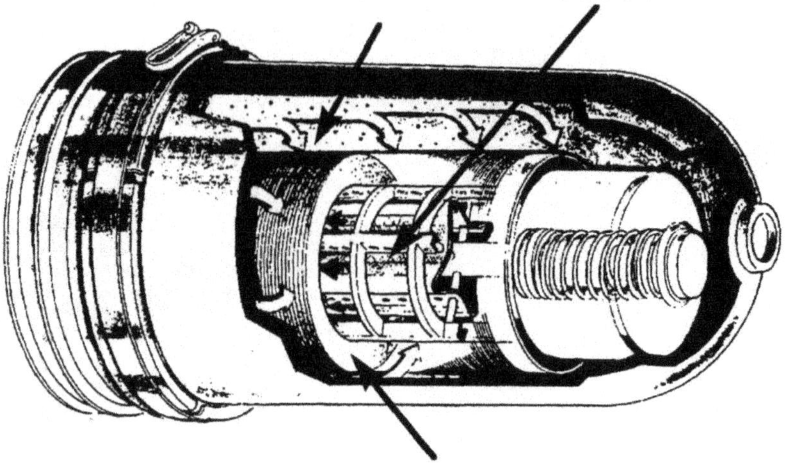

Figure 9. Disc filter.
(Netafim Irrigation, Inc.)

Automatic backflushing is available for one type of disc filter. During backflushing, clean water from a bank of filters flows into the inside of the rings and through the rings to the outside as shown in Figure 10. The rings are allowed to separate and spin during backflushing.

Smaller drip irrigation systems may use only one or two disc filters. These units must be manually cleaned by separating the rings. One manufacture makes a small filter that can be backflushed by opening a drain valve on the filter and then rotating the filter.

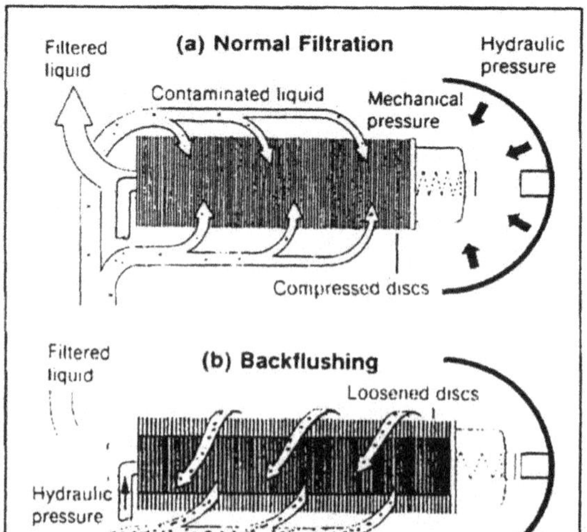

Figure 10. Disc filter in filtering and backflushing modes.
(Netafim Irrigation Inc.)

Injection Devices

By Larry Schwankl, UC Irrigation Specialist
and Blaine Hanson, UC Irrigation and Drainage Specialist

Chemigation

Chemicals are often injected through drip irrigation systems. This process, known as *chemigation,* allows a manager to apply chemicals at any time without the need for equipment in the field. Chemigation both increases the efficiency of chemical application — resulting in decreased chemical use and cost — and reduces the hazard to those handling and applying the chemicals. It is also potentially less harmful to the environment, compared, for instance, to air applications, which may cause chemical wind drift. However, chemigation can still cause environmental damage, particularly when the chemicals injected move readily with the irrigation water. Over-irrigation resulting in deep percolation can contaminate groundwater when a mobile chemical is injected.

Many different substances can be injected through irrigation systems, including chlorine, acid, fertilizers, herbicides, micronutrients, nematicides, and fungicides. Of these, chlorine, acid, and fertilizers are the substances most commonly injected. Chlorine or acid injection is used in low-volume irrigation systems to prevent clogging caused by biological growths (algae and bacterial slimes) and chemical precipitation (particularly calcium carbonate).

Irrigators wishing to inject chemicals have a variety of injection equipment from which to choose, including differential pressure or batch tanks, venturi devices, and positive displacement pumps.

Batch Tanks

Batch tanks are tanks connected to the irrigation system supply pipeline such that the irrigation water can flow into the tank and displace the chemical in the tank. The displaced chemical is forced into the irrigation supply line. As the irrigation water displaces the chemical, the injected solution becomes more and more diluted by the irrigation water as shown in Figure 1. Thus, careful management of these devices is required to insure uniform application of the chemical throughout the field.

Types of batch tank injectors are the pitot tube injector and the pressure differential injector. The pitot tube injector consists of two small open-ended tubes placed in the supply pipeline. The end of one tube faces upstream, while the end of the other tube faces downstream. Energy derived from the water velocity in the pipeline causes water to flow into the upstream tube and the chemical out of the downstream tube.

The pressure differential injector consists of inlet and outlet ports installed in the supply pipeline. A pressure reducing valve is installed between the two ports. The pressure difference causes water to flow into the upstream port, thus displacing the chemical out the downstream port.

Both types of injectors can be used where a specific amount of chemical is to be applied to a certain area. Because the chemical's concentration decreases with time, they should not be used where a desired chemical concentration is to be maintained such as injecting chlorine. Also, if more than one irrigation set is needed to irrigate the entire field, the tank should be filled prior to each set. Continuing to use the same batch over several irrigation sets will apply less chemical during the latter sets.

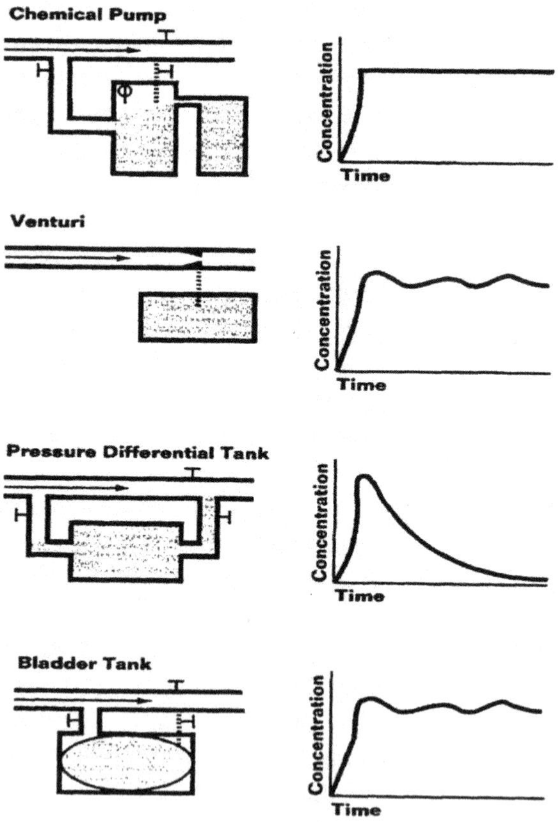

Figure 1. Chemical concentration levels during injection using a batch tank, venturi injector, and positive displacement pump.
(T-Systems International, Inc.)

Venturi Devices

Venturi devices (Figure 2) consist of a constriction in a pipe's flow area, resulting in a negative pressure or suction at the throat of the constriction. Venturi injectors are frequently referred to as "mazzei injectors." Mazzei is simply a trade name for a particular brand of venturi injector. Venturi injectors are also available from other manufacturers.

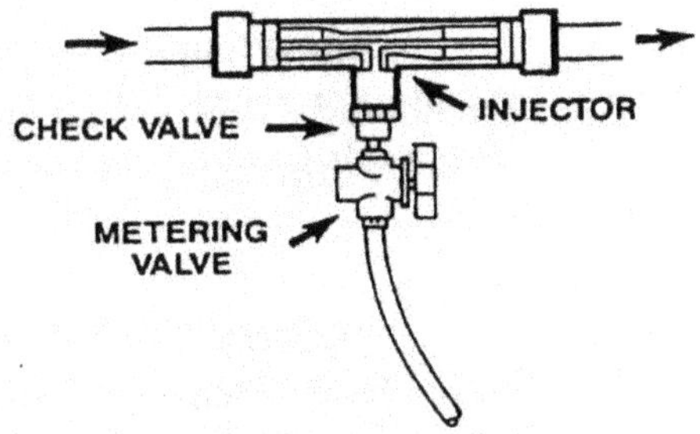

Figure 2. Venturi device.

The venturi injector is frequently installed across a valve or other point where between 10 and 30 percent of the system pressure is lost. Because of friction in the venturi, this pressure differential is necessary to allow flow through the venturi injector. This means that the inlet of the injector must be at a pressure 10 to 30 percent higher than the outlet port. Because of these significant pressure losses, the injector should be installed parallel to the pipeline so that flow through the injector can be turned off with a valve when injection is not occurring. The injection rate of a venturi device is determined by the size of the venturi and the pressure differential between inlet and outlet ports. Injection rates as high as 700 gallons per hour are possible with large venturi devices and large pressure differentials.

Figure 3 shows injection rate versus pressure drop across the venturi for several inlet pressures. Note that for each inlet pressure, as the pressure drop increases, the injection rate also increases up to some maximum. Once this maximum is reached, further increases in pressure drop do not affect the injection rate.

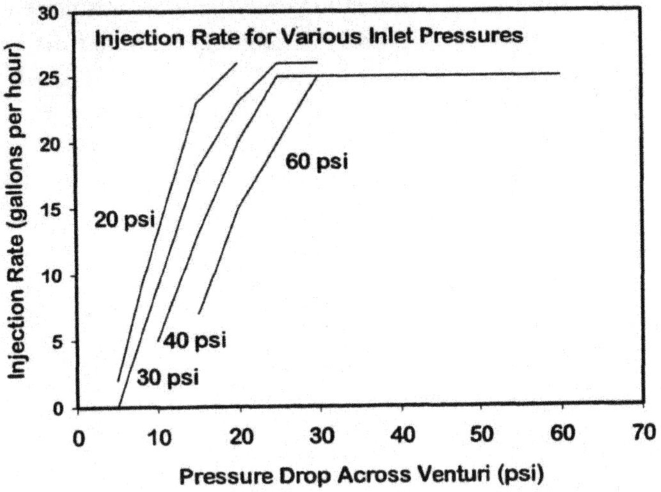

Figure 3. Venturi injection rate versus pressure drop for several inlet pressures.

A fractional-horsepower centrifugal pump is sometimes used to provide a pressure differential for venturi injectors. This eliminates the pressure drop in the mainline that occurs when a valve in the mainline is used to create a pressure drop.

Venturi devices are inexpensive and relatively simple to operate, but do not inject chemicals at as constant a rate as positive displacement pumps (Figure 1). Injecting with venturi devices may be sufficiently accurate for applications such as fertilizer injection, however.

Positive Displacement Pumps

Positive displacement pumps are piston or diaphragm pumps that inject at precise rates. The pumps are powered by either electricity or gasoline or are driven by water. The water-driven pumps can be installed in locations that lack power. When a constant and precise injection concentration is needed, positive displacement pumps are preferable (Figure 1).

Positive displacement pumps are the most expensive of the injection devices, with costs for electric pumps running $750 or more.

Considerations

Contamination can occur if: (1) the irrigation water pumping plant shuts down while the injection unit continues to operate; or (2) irrigation water is allowed to flow back through the injection unit and into the chemical storage tank causing the tank to overflow. Appropriate backflow prevention devices should be used (see chapter "Preventing Backflow").

All chemicals should be stored in separate tanks, usually made of polyethylene or fiberglass. Mixing any chemicals together should be avoided. Some mixing of chemicals, such as mixing acid and chlorine together, can produce highly toxic by products; in this case chlorine gas. Equipment made with nylon or with nylon components should be avoided. Acid injection will severely damage nylon materials.

Managing a Drip System

Irrigation Scheduling

By Blaine Hanson, UC Irrigation and Drainage Specialist

Concepts of Irrigation Scheduling

Drip irrigation can apply irrigation water very uniformly throughout a field. The more uniform the applied water, the higher the potential maximum irrigation efficiency. The uniformity of applied water depends on irrigation system design, selection of drip tape, and system maintenance.

A high irrigation efficiency requires not only good system design and maintenance, but also good management of the irrigation water. Good management means applying the right amount of water at the right time. Irrigation systems designed for high irrigation efficiencies can be inefficient if too much irrigation water is applied.

Irrigation scheduling involves determining both when to irrigate and how much water to apply. Proper irrigation scheduling ensures that the volume of water applied to the field equals the volume of water used by the crop plus water losses due to irrigation inefficiencies.

In contrast to irrigation scheduling under furrow and sprinkler irrigation, which involves determining the available soil moisture storage and calculating an allowable soil moisture depletion, irrigation scheduling under drip irrigation of row crops is a crop evapotranspiration-based approach. The soil-based approach is inappropriate for drip irrigation because local water distribution patterns and root patterns are highly variable with distance and depth from the emitter, making it almost impossible to estimate the available soil moisture. The evapotranspiration-based approach depends on estimating volume of crop evapotranspiration between irrigations, and then applying that volume of water during the irrigation. Irrigation frequencies of 1 to 3 days are commonly used for drip irrigation compared with frequencies of 10 to 14 days used for furrow and sprinkler irrigation.

Because drip irrigation wets only a small portion of the soil, it might be tempting to assume that less soil evaporation results compared with furrow irrigation. But several studies have shown that evapotranspiration is virtually the same under surface drip and furrow systems.

Under a buried drip system, however, less evaporation from the soil may occur compared with a surface drip system. The soil evaporation rate of a buried drip system is affected by the emitter discharge rate, by the soil type, and by how deeply the tape is buried. The deeper the tape is buried, the slower the evaporation rate, as soil conditions increasingly control evaporation.

Crop Evapotranspiration

The crop evapotranspiration is calculated using the reference crop ET and a crop coefficient. The crop coefficient is defined as the ratio of the crop ET to the reference crop ET. The following equation describes this relationship

$$ET_C = K_C \times ET_o$$

where ET_C is the crop evapotranspiration, K_C is the crop coefficient, and ET_o is the reference crop evapotranspiration.

Reference Crop Evapotranspiration

The reference crop normally used in California is a well-watered grass. Data on the reference crop ET are available through the CIMIS network, operated by the State Department of Water Resources.

Field experience and research shows that historical reference crop ET can be used for irrigation scheduling. The error in using historical ET instead of actual or real-time ET is generally less than 10% during the summer months. The error can increase to about 20% during the spring and fall.

Table 1 lists historical reference crop ET for selected locations in California.

Crop Coefficients

Crop coefficients are listed in Tables 2, 3 and 4 for some row crops. An alternate approach for calculating crop coefficients for row crops is to use the canopy coverage, expressed as a fraction of the width of the bed spacing. Field studies have shown that the fraction canopy coverage is a good estimate of the crop coefficient for some crops. The following equation describes this relationship.

$$K_C = \text{Width of Canopy Coverage/Bed Spacing.}$$

This approach is not valid for tree crops. The ratio of crop evapotranspiration and reference crop ET is much higher than the canopy coverage up to 60 % coverage. Crop ET remains constant with canopy coverage for coverage larger than 60 %.

Irrigation Frequency

What is an appropriate irrigation frequency under drip irrigation. Frequencies typical range from daily to once per week during periods of peak evapotranspiration. Irrigation frequencies during peak-ET periods should not exceed three days for shallow-rooted vegetable crops and should not exceed 7 days for other crops. This is discussed in the chapter "Deciding How Often to Irrigate".

Procedures

A procedure for managing drip irrigation of row crops is as follows:

1. Determine the daily reference crop evapotranspiration (ETo) using either historical or real-time values. Historical reference crop evapotranspiration is Table 1.

2. Determine the total reference crop evapotranspiration (ETto) between irrigations by multiplying the daily reference evapotranspiration by the days between irrigations, or

$$\text{ETto (inches)} = \text{ETo} \times \text{days between irrigations}$$

3. Multiply the total reference crop evapotranspiration (ETto) by the appropriate crop coefficient (Kc) to determine the total crop evapotranspiration (ETc) between irrigations, or

$$\text{ETc} = \text{Kc} \times \text{ETto}$$

4. Calculate the application rate of the drip system by

$$I = 11.55 \times q / F,$$

where I is the application rate in inches per hour, q is the tape discharge rate in gallons per minute per 100 feet of tape, and F is the furrow spacing in inches. The discharge rate of the drip system in the least-watered part of the field should be used, which occurs where the pressure is the lowest. This will ensure that all parts of the field receive an amount of water at least equal to that needed.

5. Calculate the irrigation set time needed to supply the total crop evapotranspiration by dividing the application rate into the total crop evapotranspiration, or

$$\text{Ti} = \text{ETc} / I,$$

where Ti is the irrigation set time in hours

Example

Calculate the irrigation set time needed during June 15 through 30 for tomatoes irrigated daily in the Davis area. Canopy coverage is equal to the bed spacing of 66 inches. The tape discharge rate is 0.4 gpm/100 ft.

1. Determine the daily reference crop evapotranspiration (ETo) using historical evapotranspiration:

$$\text{ETo from Table 1} = 0.29 \text{ inches per day}$$

2. Calculate the total reference crop evapotranspiration between irrigations:

ETto = 1 day x 0.29 inches per day = 0.29 inches

3. Calculate the total crop evapotranspiration between irrigations:
$K_C = 1.00$

$ET_C = 1.00 \times 0.29$ inches = 0.29 inches

4. Calculate the application rate:

I = 11.55 x 0.4 gpm/100 ft / 66 inches
= 0.07 inches per hour

5. Calculate the irrigation set time:

Ti = 0.29 inches per day/0.07 inches per hr
= 4.1 hrs per day

Using Flowmeters

Flowmeters at the irrigation pump can also be used to determine the irrigation set time. The irrigation set time can be calculated by:

Ti = (449 x A x D) / Q,

where Ti is the irrigation set time in hours, A is the number of acres irrigated, D is the depth of water to be applied in inches, and Q is the system flowrate in gallons per minute, as measured with the flow meter.

The nonuniformity of the applied water must be taken into account using this method. This requires that the total crop evapotranspiration, calculated under Step 3, needs to be divided by the field-wide distribution (DU) or emission uniformity (EU) of the irrigation system, or

D = Total Crop Evapotranspiration / (DU/100 or EU/100).

However, the minimum value of the EU used in this equation should be about 80 percent. It is not recommended that values less that about 80 percent be used, even though the actual EU or DU is below that. If the actual DU or EU is less than 80 percent, then improvements to the micro-irrigation system should be made to increase the DU or EU to at least 80 percent.

Leaching Considerations

The above procedure ensures that an amount of water equal to the crop evapotranspiration is applied in areas of the field receiving the least amount of water. An additional amount of water, above that applied to meet the crop evapotranspiration, may be needed where salinity is a problem. The additional amount is called the leaching requirement, which is that amount of water needed to maintain soil salinity at levels not injurious to the crop. The amount of water needed to meet both the crop evapotranspiration requirements and the leaching requirement can be determined from:

Amount to be applied = Crop Evapotranspiration/(1 - Lr/100),

where Lr is the leaching requirement. Some recommended leaching requirements are in the chapter "Salt Patterns Under Drip Irrigation".

Table 1. Average reference crop evapotranspiration (inches per day).

		Shafter	Five Points	Parlier	Davis	Nicolaus	Durham	McArthur	Castrovill	Santa Maria	Brawley
Jan	1-15	0.03	0.04	0.03	0.03	0.03	0.03	0.02	0.05	0.05	0.07
	16-31	0.05	0.05	0.05	0.05	0.05	0.06	0.04	0.06	0.08	0.09
Feb	1-15	0.07	0.07	0.06	0.06	0.06	0.07	0.05	0.07	0.09	0.11
	16-30	0.09	0.10	0.09	0.09	0.10	0.09	0.07	0.08	0.10	0.14
Mar	1-15	0.11	0.11	0.11	0.09	0.09	0.09	0.08	0.08	0.11	0.16
	16-31	0.14	0.15	0.13	0.14	0.12	0.12	0.11	0.11	0.14	0.19
Apr	1-15	0.19	0.20	0.17	0.18	0.15	0.16	0.15	0.13	0.17	0.22
	16-30	0.21	0.22	0.19	0.20	0.18	0.17	0.15	0.14	0.18	0.25
May	1-15	0.24	0.26	0.22	0.24	0.21	0.21	0.18	0.15	0.19	0.28
	16-31	0.26	0.27	0.24	0.25	0.21	0.22	0.20	0.15	0.19	0.29
Jun	1-15	0.27	0.31	0.26	0.28	0.24	0.25	0.22	0.16	0.20	0.31
	16-30	0.28	0.30	0.27	0.29	0.26	0.26	0.26	0.16	0.20	0.32
Jul	1-15	0.28	0.30	0.27	0.29	0.26	0.27	0.28	0.15	0.19	0.31
	16-31	0.26	0.28	0.26	0.27	0.25	0.25	0.25	0.14	0.18	0.28
Aug	1-15	0.25	0.28	0.24	0.26	0.24	0.24	0.25	0.13	0.18	0.29
	16-31	0.24	0.26	0.22	0.24	0.21	0.21	0.22	0.13	0.17	0.28
Sep	1-15	0.21	0.24	0.19	0.21	0.19	0.19	0.19	0.12	0.15	0.27
	16-30	0.18	0.20	0.15	0.18	0.16	0.17	0.15	0.10	0.15	0.22
Oct	1-15	0.16	0.17	0.13	0.16	0.13	0.14	0.12	0.10	0.12	0.19
	16-31	0.12	0.13	0.09	0.12	0.09	0.10	0.08	0.08	0.10	0.15
Nov	1-15	0.09	0.10	0.07	0.09	0.07	0.07	0.05	0.07	0.09	0.13
	16-30	0.06	0.07	0.05	0.06	0.05	0.05	0.03	0.06	0.08	0.10
Dec	1-15	0.05	0.06	0.04	0.05	0.04	0.04	0.03	0.05	0.07	0.08
	16-31	0.04	0.04	0.03	0.04	0.03	0.03	0.02	0.05	0.05	0.07

Table 2. Crop coefficients for the Sacramento Valley.

CROP COEFFICIENTS (Kc) - SACRAMENTO VALLEY												
Date	Beans (pinto) (a)	Beans (pinto) (b)	Corn (a)	Corn (b)	Corn (c)	Sugarbeets (a)	Sugarbeets (b)	Sugarbeets (c)	Tomatoes (a)	Tomatoes (b)	Tomatoes (c)	Tomatoes (d)
Jan 1-15												
16-31												
Feb 1-14						0.25						
15-28						0.25						
Mar 1-15						0.25	0.20		0.26			
16-31						0.32	0.20		0.26			
Apr 1-15			0.20			0.55	0.32	0.11	0.26	0.25		
16-30			0.20			0.77	0.58	0.22	0.28	0.25		
May 1-15	0.15		0.43	0.20		1.00	0.84	0.43	0.53	0.25	0.25	
16-31	0.22		0.71	0.20		1.10	1.12	0.65	0.80	0.44	0.25	
Jun 1-15	1.09	0.08	0.97	0.52		1.10	1.12	0.85	1.06	0.76	0.54	0.20
16-30	1.09	0.44	1.15	0.85	0.18	1.10	1.12	1.06	1.11	1.08	0.85	0.31
Jul 1-15	1.09	0.90	1.15	1.15	0.30	1.10	1.12	1.14	1.11	1.10	1.14	0.64
16-31	1.03	1.08	1.15	1.15	0.79	1.10	1.12	1.14	0.96	1.10	1.14	0.99
Aug 1-15	0.54	1.08	0.85	1.07	1.15	1.10	1.12	1.14	0.77	1.02	1.14	1.14
16-31		1.08	0.52	0.77	1.15	1.01	1.12	1.14		0.82	1.07	1.14
Sep 1-15		0.70		0.48	1.15		1.12	1.14		0.63	0.98	1.05
16-30					1.06		1.12	1.14				0.87
Oct 1-15					0.78		1.07	1.14				
16-31							0.99	1.14				
Nov 1-15								1.14				
16-30								1.14				
Dec 1-15								1.03				
16-31								0.89				
Planting	30Apr01	03Jun01	02Apr01	30Apr01	17Jun01	29Jan01	01Mar01	02Apr01	26Feb01	02Apr01	30Apr01	03Jun01
Harvesting	18Aug01	22Sep01	25Aug01	08Sep01	20Oct01	25Aug01	01Nov01	31Dec01	11Aug01	08Sep01	22Sep01	29Sep01

Table 3. Crop coefficients for the San Joaquin Valley.

CROP COEFFICIENTS (Kc) - SAN JOAQUIN VALLEY												
Date	Beans (a)	Beans (b)	Beans (c)	Corn (a)	Corn (b)	Corn (c)	Corn (d)	Corn (e)	Cotton (a)	Cotton (b)	Cotton (c)	Melons (a)
Jan 1-15												
16-31												
Feb 1-14												
15-28												0.14
Mar 1-15												0.14
16-31				0.19								0.14
Apr 1-15	0.14			0.19	0.19				0.12			0.40
16-30	0.14			0.43	0.19	0.18			0.12	0.16		0.88
May 1-15	0.46	0.14		0.76	0.44	0.20			0.21	0.16	0.19	1.10
16-31	1.11	0.42		1.10	0.76	0.48	0.19		0.50	0.28	0.19	1.10
Jun 1-15	1.15	1.12		1.17	1.05	0.75	0.21		0.78	0.60	0.52	0.89
16-30	1.15	1.12	0.13	1.17	1.17	1.01	0.55	0.31	1.05	0.91	0.84	0.35
Jul 1-15	0.93	1.12	0.39	1.08	1.17	1.10	0.88	0.67	1.20	1.18	1.15	
16-31	0.49	0.85	0.99	0.79	1.00	1.10	1.06	1.05	1.20	1.18	1.15	
Aug 1-15		0.51	1.07	0.53	0.76	1.07	1.06	1.07	1.20	1.18	1.15	
16-31			1.07		0.51	0.81	1.06	1.07	1.18	1.12	1.15	
Sep 1-15			0.88			0.56	0.90	1.05	0.82	0.92	1.06	
16-30			0.42				0.66	0.75	0.47	0.71	0.85	
Oct 1-15								0.44		0.50	0.63	
16-31											0.40	
Nov 1-15												
16-30												
Dec 1-15												
16-31												
Planting	01Apr01	01May01	16Jun01	16Mar01	01Apr01	16Apr01	16May01	16Jun01	01Apr01	16Apr01	01May01	15Feb01
Harvesting	31Jul01	15Aug01	30Sep01	15Aug01	31Aug01	15Sep01	30Sep01	15Oct01	30Sep01	15Oct01	31Oct01	30Jun01

CROP COEFFICIENTS (Kc) - SAN JOAQUIN VALLEY												
Date	Melons (b)	Melons (c)	Onions (a)	Onions (b)	Onions (c)	Potatoes (a)	Potatoes (b)	Potatoes (c)	Tomatoes (a)	Tomatoes (b)	Tomatoes (c)	Tomatoes
Jan 1-15				1.15	0.71	0.51						
16-31				1.15	0.98	0.51						
Feb 1-14				1.15	1.11	0.51	0.43					
15-28				1.15	1.11	0.51	0.43					
Mar 1-15			0.30	1.15	1.11	0.77	0.57	0.55	0.25			
16-31	0.18		0.30	1.12	1.11	1.11	0.85	0.61	0.25			0.20
Apr 1-15	0.18		0.30	1.05	1.11	1.15	1.11	0.88	0.25	0.24		0.20
16-30	0.34	0.18	0.53	0.97	1.11	1.15	1.18	1.16	0.25	0.24		0.20
May 1-15	0.72	0.18	0.83	0.90	1.11	0.88	1.18	1.21	0.46	0.24	0.25	0.52
16-31	1.11	0.49	1.14	0.82	1.11		0.92	1.19	0.80	0.52	0.28	0.87
Jun 1-15	1.11	0.81	1.14		1.11		0.57	0.87	1.12	0.77	0.51	1.00
16-30	1.11	1.10	1.14		1.06			0.55	1.16	1.03	0.74	1.00
Jul 1-15	0.78	1.10	1.04		0.86				1.07	1.12	0.97	0.95
16-31	0.29	0.91	0.92		0.64				0.92	1.06	1.12	0.85
Aug 1-15		0.36	0.80						0.77	0.92	1.08	
16-31			0.68							0.77	0.91	
Sep 1-15											0.75	
16-30				0.18								
Oct 1-15				0.20								
16-31				0.38								
Nov 1-15				0.55								
16-30				0.72	0.27							
Dec 1-15				0.88	0.27	0.51						
16-31				1.06	0.47	0.51						
Planting	16Mar01	16Apr01	01Mar01	16Sep01	16Nov01	01Dec01	01Feb01	01Mar01	01Mar01	01Apr01	01May01	23Mar01
Harvesting	31Jul01	15Aug01	31Aug01	31May02	31Jul02	15May02	15Jun01	30Jun01	15Aug01	31Aug01	15Sep01	02Aug01

Table 4. Crop coefficients for the Imperial Valley.

CROP COEFFICIENTS (Kc) - IMPERIAL VALLEY														
Date	Asparagus	Cantaloupe	Cantaloupe	Carrots	Cotton	Lettuce	Lettuce	Onions	Squash	Squash	Sugarbeets	Sugarbeets	Tomatoes	Tomatoes
		(a)	(b)			(a)	(b)		(a)	(b)	(a)	(b)	(processing)	(market)
Jan 1-15	0.25			1.06			0.76	0.75		0.45	1.12	1.10		0.45
16-31	0.25			1.06			0.83	0.75		0.53	1.06	1.10		0.45
Feb 1-14	0.25	0.42		1.06			0.82	0.75		0.92	1.01	1.10	0.41	0.45
15-28	0.25	0.42		1.06			0.72	0.79		1.30	0.96	1.10	0.41	0.52
Mar 1-15	0.25	0.50		1.00			0.61	0.88		1.07	0.90	1.10	0.43	0.69
16-31	0.25	0.70		0.92			0.50	0.98		0.76	0.84	1.10	0.73	0.87
Apr 1-15	0.45	0.88		0.85	0.40			1.01		0.47	0.78	1.10	1.01	1.04
16-30	0.78	0.96		0.78	0.40			0.78		0.18	0.73	1.10	1.20	0.98
May 1-15	0.95	0.96			0.43			0.55				1.10	1.20	0.71
16-31	0.95	0.94			0.49			0.31				1.03	1.20	0.42
Jun 1-15	0.95				0.55							0.91	1.03	
16-30	0.95				0.61							0.80	0.74	
Jul 1-15	0.95				0.66						0.18			
16-31	0.95				0.73						0.30			
Aug 1-15	0.95		0.15		0.78						0.49			
16-31	0.95		0.39		0.84						0.70			
Sep 1-15	0.95		0.61		0.86	0.17			0.19		0.89			
16-30	0.95		0.84		0.85	0.23			0.32		1.09			
Oct 1-15	0.95		0.97	0.43	0.67	0.54			0.55		1.14	0.28		
16-31	0.95		0.97	0.43	0.48	0.87			0.80		1.14	0.39		
Nov 1-15	0.95		0.73	0.57		1.02	0.30		0.85		1.14	0.64		
16-30	0.77		0.44	0.74		1.02	0.33		0.85		1.14	0.89		
Dec 1-15	0.57			0.91		0.82	0.47		0.85		1.14	1.10		
16-31	0.35			1.06		0.56	0.62		0.85		1.14	1.10		
Planting	01Jan01	31Jan01	31Jul01	30Sep01	31Mar01	31Aug01	31Oct01	31Dec01	31Aug01	31Dec01	30Jun01	30Sep01	31Jan01	31Dec01
Harvesting	31Dec01	31May01	30Nov01	30Apr02	31Oct01	31Dec01	31Mar02	31May02	31Dec01	30Apr02	30Apr02	30Jun02	30Jun01	31May02

Deciding How Often to Irrigate

Blaine Hanson, UC Irrigation and Drainage Specialist
and Stephen R. Grattan, UC Plant-Water Relations Specialist

In traditional irrigation scheduling, irrigations take place when the crop evapotranspiration—the amount of water lost through evaporation from the soil and transpiration from the plant since the last irrigation— equals the allowable soil water depletion level—the maximum amount of water that can be depleted from the soil in a particular crop without reducing yield. Irrigation should take place before the soil-water content becomes too low to supply the roots with enough water for transpiration. Because the allowable depletion level varies depending on the site, the climate, the type of crop, the crop rooting depth, and the soil type, this method of determining how often to irrigate carries with it a high degree of uncertainty.

Irrigation Scheduling Under Drip Irrigation

Under drip irrigation, much of the uncertainty of traditional irrigation scheduling can be eliminated. Since drip irrigation systems are capable of applying small amounts of water daily or several times a week without significant loss to surface runoff and deep percolation, high-frequency irrigation scheduling can be based solely on an estimate of the daily crop evapotranspiration and need not consider allowable soil water depletions. In fact, using allowable soil water depletion as a basis for determining irrigation frequency with a drip system nullifies an important advantage of using the drip method.

How Frequent Should Irrigations Be?

Under drip irrigation, irrigations should be quite frequent in order to keep soil water content at a fairly constant level. Because the root zone in drip-irrigated crops is limited to the wetted soil area, making much less of the soil water readily available to the crop, irrigation must be more frequent under drip irrigation than under furrow or sprinkler irrigation.

Research Results

Research on the irrigation frequency of vegetable crops was conducted at the University of California Westside Research and Extension Center. Onions, peppers, lettuce, and processing tomatoes were drip-irrigated at frequencies of twice per day, once per day, twice per week, and once per week. For the shallower-rooted crops (onions, peppers, and lettuce), the best yields occurred under irrigation frequencies of once per day or twice per week. Yields were reduced for the once per week frequency. Yields of the twice per day frequency were similar to those of the once per day frequency.

Figure 1 shows the pattern of soil moisture about the drip tape for the daily irrigation. This pattern was consistent throughout the irrigation period.

Extreme drying occurred at edges of the bed near the surface which would be expected. Wetting increased with depth below the surface at this distance from the drip tape. Much wetting occurred below the drip tape. Directly above the drip tape, soil moisture decreased as the depth decreased.

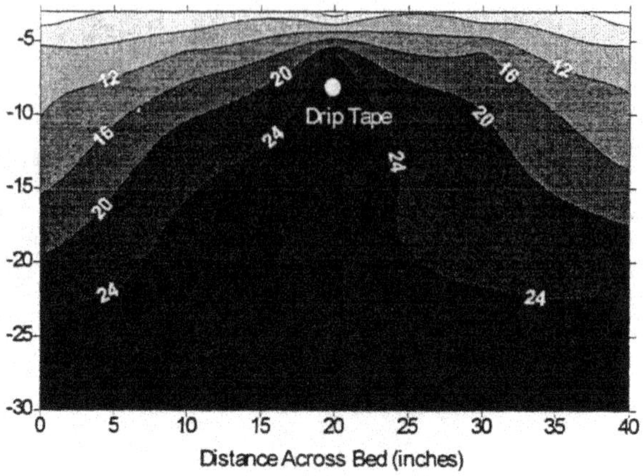

Figure 1. Soil water content under daily irrigations.
(values of contours are percent water content by volume).

Figures 2 and 3 show patterns of soil moisture before and after the once per week irrigation. Before the irrigation, in Figure 2, considerable drying occurred in the vicinity of the drip tape and less drying occurred near the edge of the bed. This suggests that most of the roots developed in the near vicinity of the drip tape. After the irrigation, soil moisture changed near the drip tape at the depth of the tape, but little change in moisture occurred above the tape and near the edges of the bed.

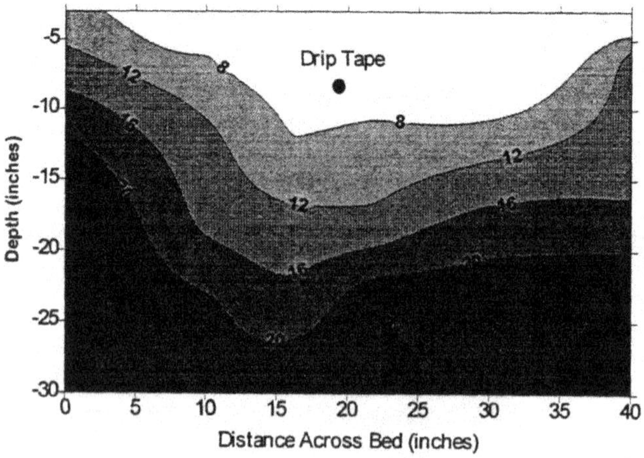

Figure 2. Soil water content just before a weekly irrigation.

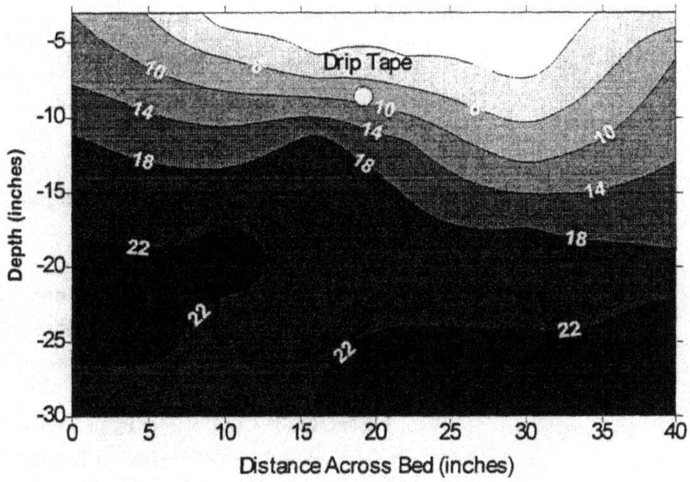

Figure 3. Soil water content just after a weekly irrigation.

An extensive review of other research on the relationship between irrigation frequency and crop yield under drip irrigation shows that irrigations can take place either several times a day, once a day, or several times a week, with little advantage of one schedule over another as long as enough water is applied each time. During consecutive hot, dry days, and where young seedlings are grown in coarse-textured soils, however, it may be best to irrigate daily until the plants are well-established, since under these conditions the plants would be extremely vulnerable to water stress.

Allowing long intervals such as weekly intervals or longer between irrigations can cause rapid depletion of the water stored in the soil particularly in sandy and loamy soils. Weekly irrigations of these soils could result in substantial percolation below the root zone because of exceeding their soil moisture storage capacity.

Scheduling multiple irrigations each day, however, can cause nonuniform water applications throughout the field because the mainlines, submains, and laterals are repeatedly filled and drained. This drainage results in more water applied to the lower part of the field than applied elsewhere.

The soil water in the root zone should be monitored continually to ensure that the roots have an adequate supply. Tensiometers are best suited to this purpose since they indicate how readily the roots can extract water from the soil. (See "Monitoring the Soil Moisture".)

References

Annual Report. 1986-1990. Water Management Research Laboratory, U.S. Department of Agriculture (ARS), Fresno, California.

Bucks. D.A., L.J. Erie, and O.F. French. 1974. Quantity and frequency of trickle and furrow irrigation for efficient cabbage production. *Agronomy Journal,* Vol. 66:53-57.

Bucks, D.A., L.J. Erie, O.F. French, F.S. Nakayama, and W.D. Pew. 1981. Subsurface trickle irrigation management with multiple cropping. *Transactions of the American Society of Agricultural Engineers,* Vol. 24:1482-89.

Davis, K.R., C.J. Phene, R. L. McCormack, R.B. Hutmacher, and D.W. Meeks. 1985. Trickle frequency and installation depth effects on tomatoes. In: *Proceedings of the Third International Drip/Trickle Irrigation Congress,* Fresno, California.

Freeman, M., J. Blackwell, and K.V. Garzolli. 1976. Irrigation frequency and total water applications with trickle and furrow systems. *Agricultural Water Management,* Vol. 1:21-31.

Hanson, B. R., D. May, and J. Egbert. 1997. Effect of irrigation frequency on drip irrigated vegetables. Report of research at the UC Westside Research and Extension Center.

Hanson, E.G., and T. C. Patterson. 1974. Vegetable production and water-use efficiency as influenced by drip, sprinkler, subsurface, and furrow irrigation methods. In: *Proceedings of the Second International Drip Irrigation Congress,* San Diego, California.

Meiri, A., H. Frenkel, and A. Mantell. 1992. Cotton response to water and salinity under sprinkler and drip irrigation. *Agronomy Journal,* Vol. 84:44-50.

Sammis, T.W. 1980. Comparison of sprinkler, trickle, subsurface, and furrow irrigation methods for row crops. *Agronomy Journal,* Vol. 72:701-4.

Monitoring the Soil Moisture

By Blaine Hanson, UC Irrigation and Drainage Specialist

For the crop to receive enough water, each irrigation must replenish the amount of water used by the crop since the last irrigation to satisfy its evapotranspiration needs. Along with estimating the crop's water requirements according to the techniques given in the preceding chapter, water managers should also monitor the soil moisture to make allowances for any conditions specific to a given site and any inaccurate assumptions made in calculating evapotranspiration.

The purpose of monitoring soil moisture is to discover any trend of decreasing or increasing soil moisture during the irrigation season. Knowledge of the actual soil moisture content is less important under high-frequency drip irrigation. If a decreasing trend is revealed, more water should be applied. If soil moisture is gradually increasing, it may be that too much water is being applied. Where irrigations are very frequent, knowing the actual soil moisture level at any given time is somewhat less important than monitoring the relative change in soil water.

Figure 1 shows tensiometer readings at the six-inch depth for drip-irrigated onions. Early in the growing season, a trend of increasing tensiometer readings with time occurred indicating insufficient irrigations. Starting in mid-May, the readings were more or less constant with time indicating adequate irrigations during that time period.

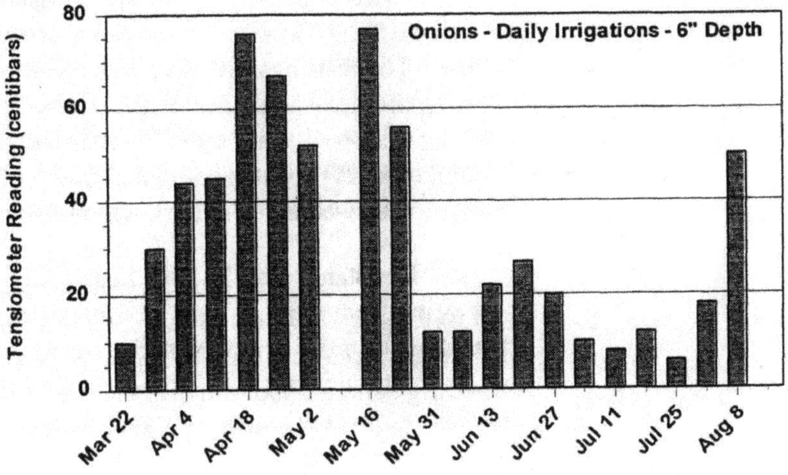

Figure 1. Tensiometer readings with time.

Tensiometers are the most practical method for growers to monitor soil moisture. Electrical resistance blocks also can be used, but some blocks are not very responsive to changes in soil moisture at high levels of soil moistures. Other instruments such as the neutron probe, capacitance soil moisture meters, or time-domain reflectometer instruments can be used, but they are expensive and sometimes difficult to use.

Tensiometers

A tensiometer consists of a 1/2" diameter plastic pipe with a porous ceramic cup attached to one end and a vacuum gauge attached to the other. The cup is saturated with water and the pipe filled with water. The cup is then inserted into the soil. As the soil repeatedly dries and wets, water flows in and out of the plastic pipe through the porous cup, producing a reading on the gauge that reflects the soil water pressure. The higher the tensiometer reading, which is usually expressed in centibars, the dryer the soil.

Place the tensiometer at about six to ten inches from the drip tape for beds with a single plant now directly above the drip tape. Place the tensiometer in the plant row for beds with two plant rows and where the drip tape is midway between the plant rows. The recommended depth of the tensiometer is about 6 to 12 inches for most crops.

Electrical Resistance Blocks

Electrical resistance blocks measure soil moisture by electrical resistance. Two electrodes are imbedded in a block of gypsum or in a ceramic block containing gypsum. The block is saturated and then installed in the soil. As the soil moisture changes, soil water flows in and out of the block, changing the block's electrical resistance. The gradual dissolution of the gypsum maintains the water in the block at a constant salinity, thereby counteracting the effect of salinity on the electrical resistance. When most of the gypsum is dissolved, the block must be replaced.

The placement of electrical resistance blocks is the same as for tensiometers. They are read with a meter that measures electrical resistance. The readings given vary depending on the particular design. One design provides a reading similar to that of a tensiometer. Field studies have shown this block, called the Watermark, to be sensitive to changes in soil moisture at high water contents. Some of the other blocks are not sensitive to soil moisture changes at high water contents.

Resistance blocks have been successively used to monitor the extent of wetting from the drip tape. By installing these instruments at various distances from the drip tape and at several depths for each distance, the wetting pattern about the drip tape can be monitored. A drying front that moves towards the drip tape may indicate insufficient applied water.

Setting the Irrigation Time

By Larry Schwankl, UC Irrigation Specialist
and Blaine Hanson, UC Irrigation and Drainage Specialist

Either of two methods can be used to calculate how long to make each irrigation so that the appropriate amount of water is applied to the field. In the first method, the irrigation time is figured from the application rate of the drip tape in gallons per minute per 100 feet, while in the second method, the time is figured using the flowrate into the field. The first method is particularly appropriate for calculating the design application rate, where the desired tape discharge rate is known. The second method is more appropriate for managing existing drip systems, where the tape discharge rate may change due to pump wear or emitter clogging and cannot be measured.

Application Rate Method

The application rate method uses the row spacing (in inches), the application rate of the drip tape (in gallons per minute per 100 feet of length), and the amount of water used by the crop between irrigations. The steps in this method are as follows:

Step 1

Step 1 is to determine the crop water use in inches per day — the standard measure for evapotranspiration-based irrigation scheduling methods.

Step 2

Step 2 is to determine the application rate of the drip tape or tubing in inches per hour (in/hr). *Table 1* can be used to make this determination if the row spacing and the irrigation system application rate (gal/min. per 100 feet) are known.

Example:

Row spacing = 60 inches
Drip tape application rate = 0.25 gal/min. per 100 ft.

From *Table 1:* the application rate = 0.05 in/hr.

Step 3

Step 3 is to determine the irrigation system operation time (in hours per day) necessary to satisfy the crop water needs. This requires the crop water use (determined in ***Step 1***), and the application rate (determined in ***Step 2***). The following formula may be used (or see *Table 2*):

$$\text{Hours of operation per day} = \frac{\text{Plant Water use (in/day)}}{\text{Application rate (in/hour)}} \qquad (1)$$

Table 1. Application rate (in/hr) of drip tapes and tubing for various flowrates and spacings.

Row spacing (in)	Flowrate (gal/min. per 100 ft) 0.1	0.15	0.2	0.25	0.3	0.35	0.4	0.45	0.5
12	0.10	0.14	0.19	0.24	0.29	0.34	0.39	0.43	0.48
14	0.08	0.12	0.17	0.21	0.25	0.29	0.33	0.37	0.41
16	0.07	0.11	0.14	0.18	0.22	0.25	0.29	0.32	0.36
18	0.06	0.10	0.13	0.16	0.19	0.22	0.26	0.29	0.32
20	0.06	0.09	0.12	0.14	0.17	0.20	0.23	0.26	0.29
22	0.05	0.08	0.11	0.13	0.16	0.18	0.21	0.24	0.26
24	0.05	0.07	0.10	0.12	0.14	0.17	0.19	0.22	0.24
26	0.04	0.07	0.09	0.11	0.13	0.16	0.18	0.20	0.22
28	0.04	0.06	0.08	0.10	0.12	0.14	0.17	0.19	0.21
30	0.04	0.06	0.08	0.10	0.12	0.13	0.15	0.17	0.19
32	0.04	0.05	0.07	0.09	0.11	0.13	0.14	0.16	0.18
34	0.03	0.05	0.07	0.08	0.10	0.12	0.14	0.15	0.17
36	0.03	0.05	0.06	0.08	0.10	0.11	0.13	0.14	0.16
38	0.03	0.05	0.06	0.08	0.09	0.11	0.12	0.14	0.15
40	0.03	0.04	0.06	0.07	0.09	0.10	0.12	0.13	0.14
42	0.03	0.04	0.06	0.07	0.08	0.10	0.11	0.12	0.14
44	0.03	0.04	0.05	0.07	0.08	0.09	0.11	0.12	0.13
46	0.03	0.04	0.05	0.06	0.08	0.09	0.10	0.11	0.13
48	0.02	0.04	0.05	0.06	0.07	0.08	0.10	0.11	0.12
50	0.02	0.03	0.05	0.06	0.07	0.08	0.09	0.10	0.12
52	0.02	0.03	0.04	0.06	0.07	0.08	0.09	0.10	0.11
54	0.02	0.03	0.04	0.05	0.06	0.07	0.09	0.10	0.11
56	0.02	0.03	0.04	0.05	0.06	0.07	0.08	0.09	0.10
58	0.02	0.03	0.04	0.05	0.06	0.07	0.08	0.09	0.10
60	0.02	0.03	0.04	0.05	0.06	0.07	0.08	0.09	0.10

Example

Crop water use = 0.3 in/day
System application rate = 0.05 in/hour

$$\text{Hours of operation per day} = \frac{0.3 \text{ in/day}}{0.05 \text{ in/hour}} = 6 \text{ hrs/day/set}$$

Using *Table 2* gives the same operation time for this example.

Table 2. Operation time (hrs/day) for various application rates (in/hr) and crop water use (in/day).

		Application rate (in/hr)									
		0.05	0.1	0.15	0.2	0.25	0.3	0.35	0.4	0.45	0.5
Plant water use (in/day)	.05	1.0	0.5	0.3	0.3	0.2	0.2	0.1	0.1	0.1	0.1
	0.1	2.0	1.0	0.7	0.5	0.4	0.3	0.3	0.3	0.2	0.2
	0.15	3.0	1.5	1.0	0.8	0.6	0.5	0.4	0.4	0.3	0.3
	0.2	4.0	2.0	1.3	1.0	0.8	0.7	0.6	0.5	0.4	0.4
	0.25	5.0	2.5	1.7	1.3	1.0	0.8	0.7	0.6	0.6	0.5
	0.3	6.0	3.0	2.0	1.5	1.2	1.0	0.9	0.8	0.7	0.6
	0.35	7.0	3.5	2.3	1.8	1.4	1.2	1.0	0.9	0.8	0.7
	0.4	8.0	4.0	2.7	2.0	1.6	1.3	1.1	1.0	0.9	0.8

Field Flowrate Method

The field flowrate method of calculating the irrigation set time requires knowing the flowrate into the field in gallons per minute, the total number of acres to be irrigated, the number of sets per irrigation over the total acreage, and the crop water use between irrigations. The steps to be used are as follows:

Step 1

Step 1 is to calculate the time required to irrigate the total acres using the following equation:

$$T = \frac{449 \times D \times A}{Q}$$

where T = the number of hours needed to irrigate the field
A = acres irrigated
Q = flowrate into the field in gallons per minute
D = depth of water to be applied, in inches

Step 2

Step 2 is to calculate the irrigation set time by dividing the number of sets into the time T.

Calculate the daily irrigation set time for a 40-acre field irrigated at a flowrate of 300 gallons per minute. Daily crop water use is 0.3 inches per day and three sets per day are used.

Step 1: $T = \frac{449 \times 0.3 \text{ inches per day} \times 40 \text{ acres}}{300 \text{ gpm}}$

$= 18$ hours

Step 2: Set time $= \frac{18 \text{ hours}}{3 \text{ sets}}$

$= 6$ hours per set

Soil Water Patterns Under Drip Irrigation

By Blaine Hanson, UC Irrigation and Drainage Specialist

The pattern of soil water content under drip irrigation is very different from soil water content patterns under other irrigation methods. In surface and sprinkler irrigation, most of the soil surface is wetted during the irrigation, resulting in wetting throughout the soil profile, but under drip irrigation, the soil water content varies throughout the soil profile, with the water content highest near the emitter and lowest midway between emitters.

Single Drip Tape

Figure 1 shows the wetting patterns about a buried drip tape used for irrigating tomatoes. Depth of the drip tape was about eight inches and soil type was a loam. Soil water content is high near the drip tape and gradually decreases with distance and depth below about 33 inches. Soil water content is the smallest near the edges of the bed.

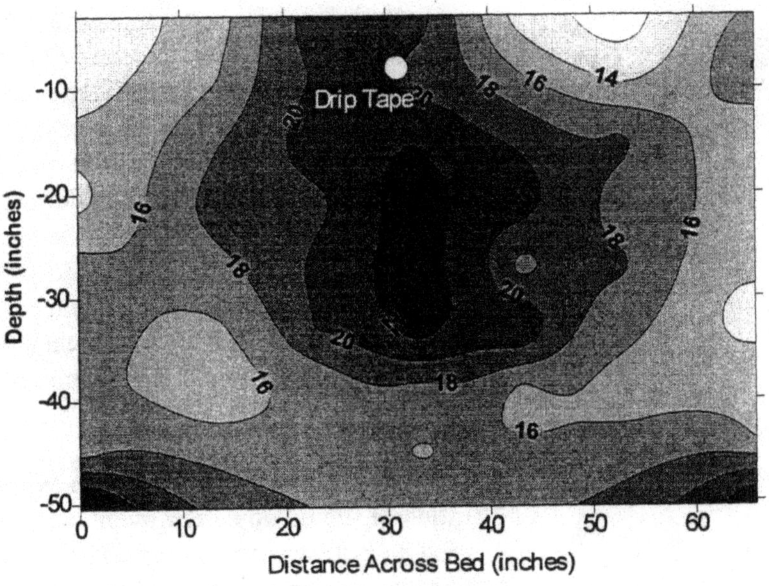

Figure 1. Pattern of soil water content under tomatoes.
(Values of soil water content are in percent on a volume basis.)

Two Drip Tapes

Figure 2 shows the wetting pattern of drip irrigation of strawberries. Two drip laterals per bed are used. Tape depth is about three inches installed in a sandy soil. Again, soil water content is highest near the emitters and gradually decreases with distance and depth.

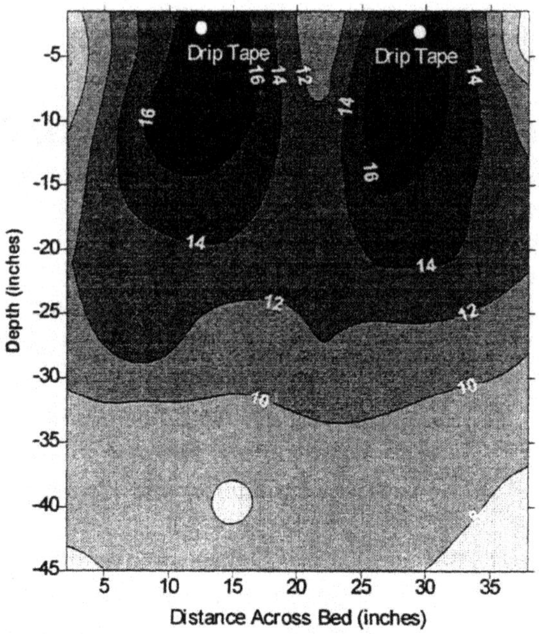

Figure 2. Wetting pattern under two drip laterals in a bed of strawberries.
(Value are percent of water control on a weight basis.)

Wetting in a Clay Soil

Under deficit irrigation, the pattern of soil water content shown in Figure 3 can occur. At this site, deficit irrigation of processing tomatoes occurred near the end of the growing season to improve the solids content. In June, the wetting pattern prior to imposing deficit irrigation, shows highest water contents near the surface with a slight decrease with depth. In August, the wetting pattern after 60 days of deficit irrigation, shows highest soil water content within the top 18 inches of the profile. A large decrease in soil water content occurred below 18 inches. The near-surface soil water content, slightly smaller than the earlier soil water content, reflect the drip irrigation. At the deeper depths, however, substantial water depletion occurred. The deficit irrigation apparently encouraged root development at depth down to at least 60 inches.

Wetting problems have been experienced in some clay soils where little lateral and upward movement were found to occur. Figure 4 shows the wetting patterns, where the tape depth was 10 inches and the application rate was about equal to the crop evapotranspiration. This pattern was developed with measurements from electrical resistance blocks. Little lateral wetting appeared to occur beyond six inches from the drip tape. When the applied water was only about 50% of the evapotranspiration, severe water depletion took place throughout the root zone, resulting in the water pattern shown in Figure 5. A severe depleting of soil water occurred throughout the profile. The water patterns shown in Figure 6 occurred when the applied water exceeded the evapotranspiration by about 50 percent. Little lateral and upward movement occurred in this soil even though the applied water exceeded the evapotranspiration.

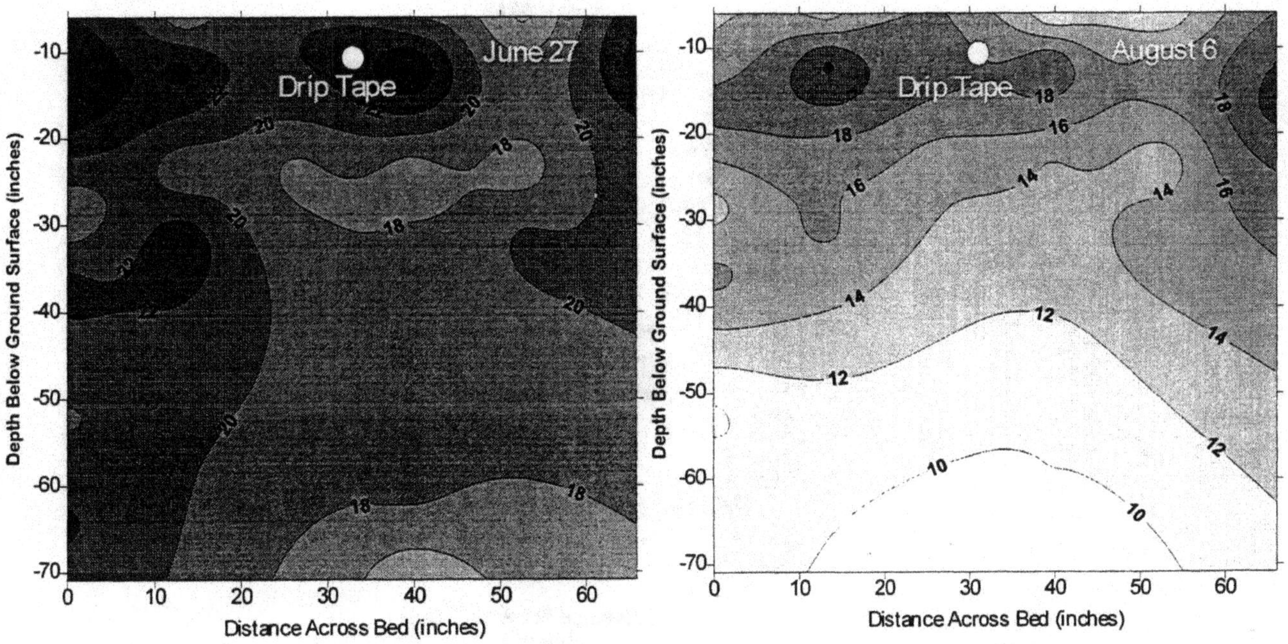

Figure 3. Patterns of soil water content under deficit irrigation during the later part of the crop season.

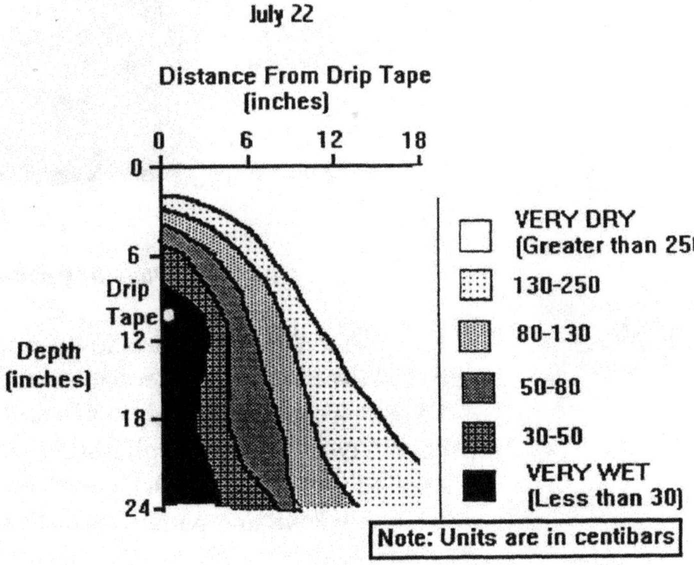

Figure 4. Pattern of soil water content on July 22.
(Values are expressed as soil water tension in centibars.)

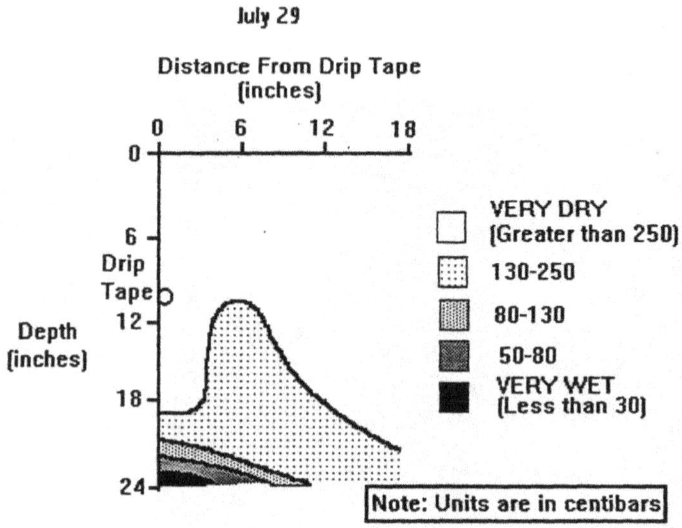

Figure 5. Pattern of soil water content on July 29.

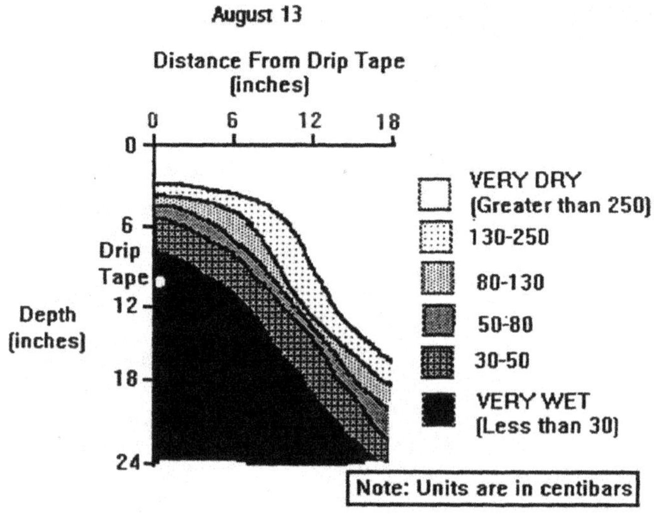

Figure 6. Pattern of soil water content on August 13.

Deep Percolation

Deep percolation below the drip tape is inherent with drip irrigation because the wetting pattern about the drip tape is uneven and because water flows more readily downhill than uphill or laterally. Even when water applications equal evapotranspiration-which would suggest 100 percent irrigation efficiency-the actual efficiency will be less than 100 percent because the uneven soil wetting pattern causes deep percolation. Decreasing the amount of water applied per irrigation to an amount less than evapotranspiration can cut down on deep percolation, but may also reduce yield.

Summary

These studies on wetting patterns about the drip tape showed better lateral movement of water in medium textured soils than in both coarse and fine-textured soils. One study, however, showed more lateral movement in a clay loam soil using low-flow drip tape compared with high-flow drip tape. The longer on-time with the low-flow tape appeared to be responsible for this behavior.

Salt Patterns Under Drip Irrigation

By Blaine Hanson, UC Irrigation and Drainage Specialist

Salt movement is governed by water movement. Under drip irrigation, water moves in a more or less radial pattern around the emitter, and salinity in the soil eventually mirrors this pattern.

Salinity Patterns After Irrigation

Figure 1 shows salinity patterns for leaching fractions of 5% and 25% under surface drip irrigation. (The leaching fraction is applied water that exceeds the evapotranspiration since the last irrigation and that therefore percolates down past the root zone as excess water.) The following can be concluded from these patterns:

- Salinity is lowest directly beneath the plant row and emitter. This low salinity zone is greatest for the high leaching fraction and least for the low leaching fraction.

- Salinity gradually increases as the distance from the emitter increases. The increase is smallest in the vertical direction and largest in the horizontal direction. With the low leaching fraction, levels of increased salinity occur closer to the emitter.

- Salinity is highest midway between emitters. This zone is smallest for the high fraction and largest for the low leaching fraction. At the midway point, salinity decreases as the depth increases.

These salt patterns reflect water movement during and between irrigations. During irrigations, salt leaching occurs in the vicinity of the emitter. The infiltrating water carries these leached salts away from the emitter. As the horizontal distance from the emitter increases, soil salinity increases because the amount of leaching decreases. Salt accumulation is highest midway between emitters because no leaching occurs in those areas.

When rainfall or water applied with another irrigation system infiltrates the soil, salts accumulated near the surface are carried downward with the infiltrating water. Leaching can therefore occur in areas with high salt accumulations, thereby lowering the salinity near the surface. The leaching process moves salt downward as a front or slug, as shown in Figure 1. Figure 1 also shows salt distribution after rainfall.

Effect on Plant Growth

Concerns have been raised about the effect on crop growth of this salt front moving downward. Figure 2 shows a distribution of soil salinity and root density. Again, soil salinity is highest near the surface. If rainfall occurs, these salts will move downward, but most of the plant roots are located near the emitter, where salinity is lowest, with substantially fewer roots growing in areas of maximum salinity and in the soil profile immediately below those areas. During a rainfall, therefore, salts near the soil surface will be leached downward into areas with relatively low root densities. This suggests that the salt movement will have little effect on crop growth.

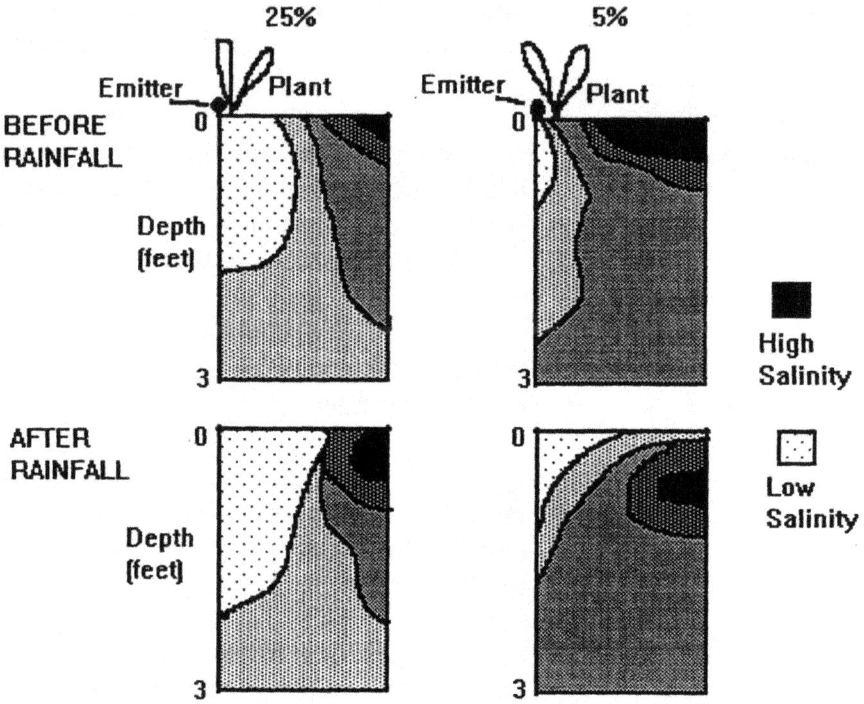

Figure 1. Salt patterns under surface drip irrigation.

In Figure 1, the emitter and plant are at the same location. Other data on salinity distributions under trickle irrigation show that as the emitter is moved away from the plant, the high salinity zone can shift toward the plant, and in some cases can occur at the plant within the active root zone.

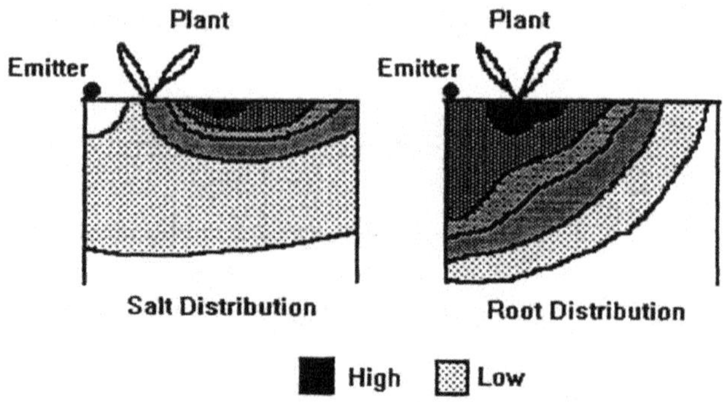

Figure 2. Distribution of soil salinity and root density.

Subsurface Drip Irrigation

The salt patterns that form under buried drip irrigation are different from those that form under surface drip irrigation. Figure 3, which illustrates a salt pattern occurring under subsurface drip irrigation, shows very high soil salinity levels near the ground surface and extending through the top few inches of the soil surface. These salinity levels can exceed 10 dS/m, depending on the salinity of the irrigation water. Salinity decreases with depth through the soil profile and increases with horizontal distance from the emitter. Near the drip tape, soil salinity is relatively low, and directly beneath the drip tape, soil salinity changes only slightly with depth.

This salt pattern shows that drip irrigation does not cause salt leaching above the drip tape, but does cause substantial leaching beneath the tape and in its immediate vicinity. Leaching diminishes under the tape as horizontal distance increases.

Salt Control

Since drip irrigation does not accomplish leaching above the drip tape, leaching will have to be done with another irrigation system or through rainfall. If rainfall is used for leaching, the drip system should be operated to replenish the soil water content to field capacity and to increase the leaching effectiveness of the rainfall, since no leaching will take place until the soil moisture exceeds field capacity. Figure 4 shows the salt pattern after about 14 inches of rainfall.

To lessen salinity just before planting, the bed of the furrow can be built up and the drip system operated to accumulate the salt in the built-up portion of the bed. The top of the bed can then be removed before planting to leave a relatively low-salt seedbed.

Subsurface drip irrigation can present a salinity hazard if rainfall during the crop season moves a slug of very high salt concentration down into the root zone, to the detriment of shallow-rooted crops. To minimize salinity damage, the drip system can be operated during the rainfall to dilute the salt.

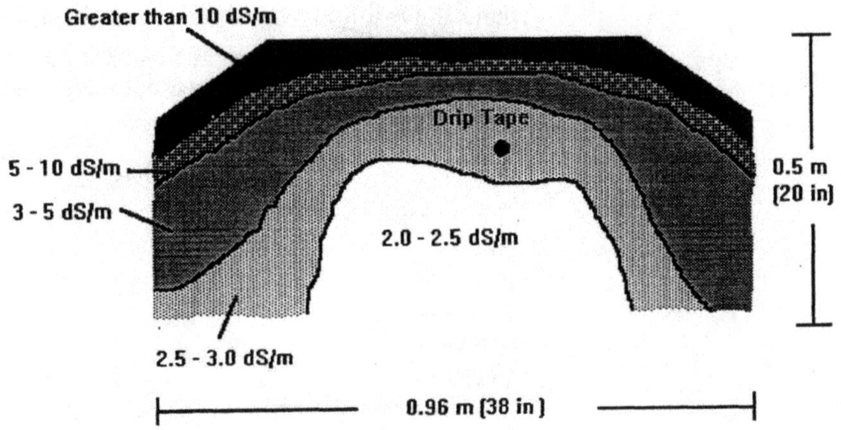

Figure 3. Salt pattern under subsurface drip irrigation.

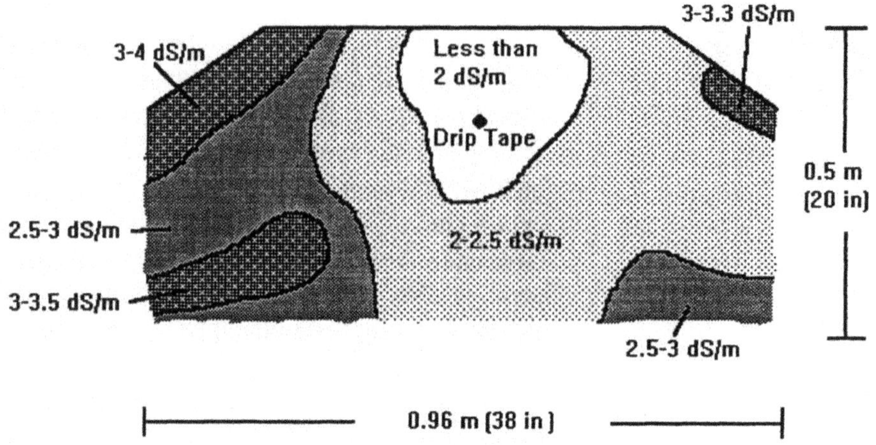

Figure 4. Salt pattern after 14 inches of rainfall.

Leaching Fractions Under Drip Irrigation

Leaching fractions function differently under drip irrigation than in traditional irrigation practices. With conventional methods, leaching fractions result in an average soil salinity. But under drip irrigation, where leaching ranges from very small amounts midway between laterals to very large amounts beneath the emitter, the leaching fraction results in an area of low salinity soil adjacent to the emitter. The greater the leaching fraction, the greater the volume of low salinity soil.

How much leaching is necessary with drip irrigation? Table 1 list some leaching requirements for vegetable crops for irrigation water salinity of 1 and 2 dS/m. These requirements are based on research by the U.S. Salinity Laboratory.

The amount of applied water needed to satisfy both ET and leaching can be calculated by:

$$\text{Applied water} = \frac{\text{Crop Evapotranspiration}}{(1\text{-Leaching Requirement})}$$

where the leaching requirement is expressed as a decimal fraction.

Table 1. Recommended leaching requirements for some vegetable crops.

Crop	Leaching Requirement (%)	
	1 dS/m	2 dS/m
Lettuce	15	26
Celery	10	14
Cauliflower	10	17
Tomatoes	12	21
Strawberry	18	26

References

Hoffman G.J., S.L. Rawlins, J.D. Oster, and S.D. Merrill. 1979. Leaching requirement for salinity control I. Wheat, sorghum, and lettuce. *Agricultural Water Management,* Vol. 2: 177-92.

Hoffman, G.J., and J.A. Jobes. 1983. Leaching requirement for salinity control III. Barley, cowpea, and celery. *Agricultural Water Management,* Vol. 6:1-14.

Hoffman, G. J. , M. C. Shannon, and J.A. Jobes. 1985. Influence of rain on soil salinity and lettuce yield. In *Irrigation in Action,* Proceedings of the Third International Drip/Trickle Irrigation Congress, November 18-21, 1985, Fresno, CA.

Jobes, J.A., G. J. Hoffman, and J.D. Wood. 1981. Leaching requirement for salinity control II. Oat, tomato, and cauliflower. *Agricultural Water Management,* Vol. 4: 393-407.

Moshrefi, N. and F. Beese. 1985. Effect of irrigation system on salt and root distribution. In *Irrigation In Action,* Proceedings of the Third International Drip/ Trickle Irrigation Congress, November 18-21, 1985, Fresno, California.

Plaut, Z., M. Rom, and A. Meiri. 1985. Effect of irrigation system on salt and root distribution. In *Irrigation In Action,* Proceedings of the Third International Drip/Trickle Irrigation Congress, November 18-21, 1985, Fresno, California.

Yaron, B., J. Shalhevet, and D. Shimshi. 1973. Patterns of salt distribution under trickle irrigation. In *Physics of Soil Water and Salt,* Ecological Studies 4. Springer-Verlag.

Germinating Seeds

By Blaine Hanson, UC Irrigation and Drainage Specialist

How well seeds germinate depends partly on the water content of the seed bed. In buried drip irrigation this is determined by water moving upward from the drip tape, which is influenced by the emitter spacing, the soil type, any residual water content from rain, how much water is applied, and how deep the tape is buried.

Very few studies have been done on seed germination under buried drip irrigation, but some findings have been reported:

• In an initially dry sandy loam soil, drip tape buried at 8 inches resulted in a high degree of seedling emergence in cotton, but emergence was poor when the tape was buried at 18 inches. Emitter spacing of 12 inches resulted in better emergence than emitter spacing of 24 inches. When the soil contained more residual rainwater at planting, all tape depths produced good stands, although lower depths delayed emergence.

• In tomatoes grown in a clay loam soil, when seed depth was 1.5 inches, emergence was best when drip tape was buried at 6 and 9 inches. Emergence was less favorable when the tape was buried at 12 inches. A seed depth of 2.5 inches resulted in poor emergence. The deeper the seed depth, the slower the emergence.

• Cotton germinated in a fine sand down to tape depths of 8 inches. The deeper the tape, the longer the period between planting and emergence.

These studies, grower experience, and the other studies listed in the references on the following page suggest that with emitter spacings of 12 inches, seeds will germinate adequately at tape depths of 8 inches or less provided a good seed bed can be established. One grower has reported good emergence with a tape depth of 12 inches and an emitter spacing of 16 inches, but another grower found that a tape buried at 10 inches in a clay loam soil with emitter spacings of 24 inches failed to wet the soil surface. These results indicate that if the drip system is to be used for seed germination, the tape generally should be buried less than 8 inches deep and the emitters should be spaced between 12 and 16 inches apart. If there is no residual water in the soil at the time of planting and the tape is deeper than 8 inches, sprinkling may be necessary to germinate seeds.

The salinity of the soil can also affect seed germination. In saline conditions, soil salinity may be extremely high in the top one or two inches of the soil. Leaching the salts that have accumulated in this interval may be necessary for germination.

References

Constable, G.A., I. J. Rochester, and A. S. Hodgson. 1990. A comparison of drip and furrow-irrigated cotton on a cracking clay soil: yield and quality of four cultivars. *Irrigation Science,* Vol. 11:149-53.

DeTar, W.R., and L. R. Elliott. 1990. Field testing of various subsurface dripperlines for use with cotton an a very sandy soil. *1990 Annual Report,* U.S. Department of Agriculture Cotton Research Station, Shafter, CA; pp. 21-23.

Hanson, E.G., and T.C. Patterson. 1974. Vegetable production and water-use efficiency as influenced by drip, sprinkler, subsurface, and furrow irrigation methods. *Proceedings, Second International Drip Irrigation Congress,* San Diego, CA, July 7-14, 1974.

Plaut, Z., M. Rom, and A. Meiri. 1985. Cotton response to subsurface trickle irrigation. *Drip/Trickle Irrigation in Action. Proceedings of the Third International Drip/Trickle Irrigation Congress,* Fresno, CA, November 18-21, 1985.

Sammis, T.W. 1980. Comparison of sprinkler, trickle, subsurface, and furrow irrigation methods for row crops. *Agronomy Journal,* Vol. 72: 701-4.

Schwankl, L., S.R. Grattan, and G. M. Miyao. 1991. Drip system design, management promote seed emergence. *California Agriculture,* Vol. 45:21-23.

Willardson, L. S. 1974. Seed germination response to subsurface irrigation depth. *Proceedings, Second International Drip Irrigation Congress,* San Diego, CA July 7-14, 1974.

Root Patterns Under Drip Irrigation

By Blaine Hanson, UC Irrigation and Drainage Specialist

How roots develop in row crops irrigated with a drip system depends on the type of crop, the wetting pattern, the location of the drip tape relative to the plant row, the salinity of the water, and how the drip system is managed. The type of crop determines the potential root pattern—that is, whether the crop will be shallow-rooted or deep-rooted—but within each plant type, the root pattern is determined mainly by the soil water content, which is determined in turn by the wetting pattern of water flowing from the drip tape, by rainfall supplementing the drip irrigation, and by any variations in the soil.

The figures below illustrate root development patterns under various conditions. Figures 1 and 2 show root patterns in sweet corn under a surface drip tape and under a drip tape buried at 12 inches, where drip irrigation is the sole water source and where the emitter is located along the plant row.

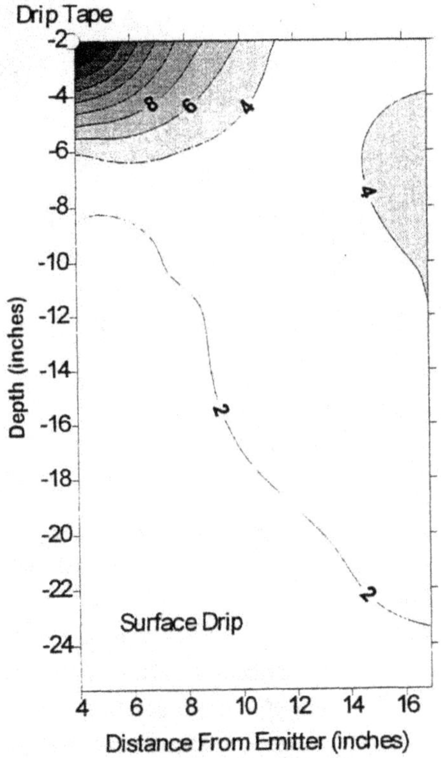

Figure 1. Root pattern in sweet corn under surface drip irrigation.
(Contour values are in grams of root weight.)

Figure 1 illustrates differences in root density developing under surface drip irrigation depending on the horizontal and vertical distance from the emitter. Root density is greatest within six inches of the drip lateral and within four inches beneath the emitter. Between six and ten inches laterally from the emitter and extending four to eight inches deep, root density was about 7 percent of the total root mass. Beyond those distances root growth was minimal.

Figure 2 shows the pattern of root development in sweet corn under drip tape buried 12 inches deep. Here the root density is greatest (nearly 26 percent of the total mass) at the 12-inch depth just above the emitter and gradually decreases to about 18 percent of the total root mass in the first four inches above the emitter. Between six to ten inches from the emitter, the root mass amounted to between 4 and 5 percent of the total mass.

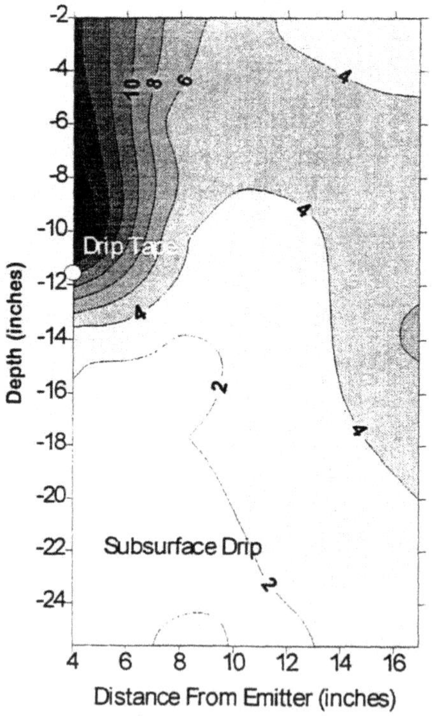

Figure 2. Root pattern in sweet corn under a drip tape buried at 12 inches.
(Contour values are in grams of root weight.)

Where drip irrigation supplements rainfall, a different root pattern may result. One study of cotton irrigated with a drip system buried at various depths showed that at up to 20 inches laterally from the emitter root patterns were the same regardless of tape depth, with root density highest near the soil surface, reflecting stored soil moisture from rainfall.

The patterns shown in Figures 1 and 2 reflect root patterns resulting when the emitter is placed along the plant row, but when the emitter or lateral is offset from the plant row, the roots tend to grow toward the emitter, so long as the distance from the emitter to the plant is not so great that little water from the emitter reaches the plant row.

References

Martinez, Hernandez, J. J., B. Bar-Yosef, and U. Kafkafi. 1991. Effect of surface and subsurface drip fertigation on sweet corn rooting, uptake, dry matter production and yield. *Irrigation Science*, Vol. 12:153-59.

Kamara, L., R. Zartman, and R.H. Ramsey. 1991. Cotton-root distribution as a function of trickle irrigation emitter depth. *Irrigation Science*, Vol. 12:141-44.

Batchelor, C.H., G.C. Soopramanien, and J.P. Bell. 1985. Root development of drip-irrigated sugar cane. *Proceedings of the Third International Drip/Trickle Irrigation Congress, Drip/Trickle Irrigation In Action*, Vol. 2:687-95. Fresno, California, November 18-21, 1987.

Phene, C.J., K.R. Davis, R.B. Hutmacher, B. Bar-Yosef, D.W. Meeks, and J. Misaki. 1991. Effect of high frequency surface and subsurface drip irrigation on root distribution of sweet corn. *Irrigation Science*, Vol. 12:135-40.

Controlling Phytophthora Root Rot

By Stephen R. Grattan, UC Plant-Water Relations Specialist

Because irrigations are more frequent in drip irrigation than in conventional irrigation methods, and because wetting patterns are different, measures to control disease must also differ when drip irrigation is used.

While research is still preliminary, studies so far indicate that if the right techniques are used, drip irrigation can effectively control outbreaks of *Phytophthora* root rot — a disease that has caused widespread yield loss in California row crops such as peppers, tomatoes, and squash. Absent careful management, though, drip irrigation can create the very soil conditions that promote *Phytophthora* root rot. Genetic resistance and chemical control methods have not been effective in controlling this disease.

Phytophthora root rot thrives where soil remains in a water-saturated state for long periods of time. How frequently irrigations are performed, how much water is applied each time, and — in drip irrigation — where the drip tapes are placed, all bear on the volume and configuration of the saturated soil and therefore on the potential for a disease outbreak.

Since infection is substantially reduced where soil water is only slightly below the saturation point, shifting this saturated soil zone away from the stems and major roots of the plant by the placement of the tape and by the rate at which water is delivered, can greatly lessen disease potential. It appears that subsurface drip irrigation, where drip tape is buried 8 to 12 inches beneath the soil surface, can accomplish this purpose and that this irrigation method is more likely than furrow irrigation methods to limit *Phytophthora* root rot.

Field experiments conducted at UC Davis where *Phytophthora* root rot was incorporated into the crop root zone, found that drip irrigation lowered the incidence of disease in tomatoes and peppers compared to furrow irrigation. The placement of the tape proved to be significant. *Phytophthora* root rot was more prevalent when the tape was buried within 3 inches of the soil surface than when it was buried deeper — up to 12 inches beneath the surface.

In peppers, there was a lower incidence of disease with a subsurface drip tape buried at 8 inches than in a furrow-irrigated crop.

These tests showed no consistent relationship between disease incidence and how often irrigation took place—from several times a day to every other day. But the tests did find that increasing the amount of water applied to one and a half times evapotranspiration increased disease.

In citrus, *Phytophthora* root rot was more severe under surface drip irrigation than under furrow, probably because drip irrigation applied water more often. Surface drip irrigation also increased the disease in peppers grown in moderate rainfall conditions.

While the experiments conducted so far indicate promise in the potential for controlling *Phytophthora* root rot with drip irrigation, specific guidelines for managing drip irrigation to control this disease will have to await the results of further study.

References

Cafe-Filho, A.C., and J.M. Duniway. 1993. Management of *Phytophthora* root rot of pepper by placement of drip irrigation emitters. *Biological and Cultural Tests for Control of Plant Diseases,* APS Press. Vol. 8:27.

Duniway, J.M. 1983. Role of physical factors in the development of *Phytophthora* diseases. In: D.C. Erwin, S. Bartnick-Garcia, and P.H. Tsao, Eds. *Phytophthora: Its biology, taxonomy, ecology, and pathology.* APS Press, St. Paul, MN, pp. 175-87.

Duniway, J.M., and S.R. Grattan. 1990. Unpublished results.

Feld, S. J., and J.A. Menge. 1990. Influence of drip and furrow irrigation on *Phytophthora* root rot of citrus under field and greenhouse conditions. *Plant Disease.* Vol. 74:21-27.

Ristaino, J.B. 1991. Influence of rainfall, drip irrigation, and inoculum density on the development of *Phytophthora* root and crown rot epidemics and yield in bell pepper. *Phytopathology,* Vol. 81: 922-29.

Weed Management

By Stephen R. Grattan, UC Plant-Water Relations Specialist

Drip irrigation makes it possible to control weeds with much less use of pre-emergent herbicides than with furrow or sprinkler irrigation methods. With a buried drip system, weed growth is restricted to a narrow strip of soil above the drip tape, since only that portion of the soil surface becomes wet. So long as the weather is relatively dry—as is the case with most summer annual crops grown in the Central Valley and desert regions of California — furrows will remain weed-free.

In one study of processing tomatoes, subsurface drip tubing buried 10 inches below the soil surface was as effective at controlling annual weeds (redroot pigweed and barnyard grass) as herbicides (napropamide and pebulate) used with sprinkler and furrow irrigation.

Herbicides are also likely to be less effective with drip irrigation than with furrow or sprinkler irrigation, since many pre-emergent herbicides break down more quickly when soil moisture remains high for long periods of time than under drier conditions. In conventional irrigation, the interval between irrigations is usually between 7 and 21 days, while under drip irrigation, irrigation takes place daily or two to three times a week, creating the moist soil conditions that can cause these herbicides to prematurely degrade.

More field research is needed to test various weed control options in row crops irrigated with drip systems, but indications are that drip irrigation can not only help reduce herbicide use, but may also facilitate the use of other weed control techniques impractical under furrow or sprinkler irrigation.

References

Fisher, B.B., D.A. Goldhamer, T. Babb, and R. Kjelgren. 1985. Weed control under drip and low-volume sprinkler irrigation. *California Agriculture,* Vol. 39:24-25.

Grattan. S.R., L.J. Schwankl, and W.T. Lanini. 1988. Weed control by subsurface drip irrigation. *California Agriculture,* Vol. 42:22-24.

Fertilizing With A Drip System

Using a Drip System to Apply Fertilizer: An Overview

By Larry Schwankl, UC Irrigation Specialist
and Terry Prichard, UC Water Management Specialist

With drip irrigation, fertilizer can be injected through the irrigation system. Drip irrigation systems are well suited to this process — called *fertigation* — because irrigations are frequent and because the amount of water applied can be easily controlled. Applying fertilizers through a drip system:

- makes fertilizer distribution relatively uniform
- allows flexibility in timing fertilizer application
- reduces the labor required for applying fertilizer compared to other methods
- allows less fertilizer to be used compared to other fertilization methods
- offers the potential for lowering costs

Contamination Hazard

Measures must be taken, however, to guard against contaminating the soil, the chemical supply, or the water supply. Contamination can occur if (1) the irrigation water pumping plant shuts down while the injection equipment continues to operate, causing contamination of the water source or unnecessary amounts of fertilizer to be injected into the irrigation system, or (2) the injection equipment stops while the irrigation system continues to operate, causing the irrigation water to flow into the chemical supply tank and to overflow onto the ground.

Preventing Backflow

Several types of devices are available for preventing backflow, including vacuum breakers (atmospheric and pressure types) and check values (single and double) are available. Local regulations should be followed in selecting and using these devices.

If the injection pump is electrically driven, an interlock should be installed so that the injection pump will stop if the irrigation system pump shuts down. To keep water from flowing backward into the chemical tank, a check valve or a solenoid valve, normally kept closed, can be installed in the injection line following the injector. If an electrical solenoid valve is used, it should be connected to the injector pump and interlocked with the irrigation pump. (See the chapter, "Preventing Backflow", for more information.)

Fertilizer Solubility

In order to be injected, fertilizers must be soluble. Fertilizers delivered as a solution can be injected directly into the irrigation system, while those in a dry granular or crystalline form must be mixed with water to form a solution. Fertilizer materials differ widely in water solubility, with solubility depending on the physical properties of the fertilizer as well as on irrigation water temperature and pH. Dry fertilizers are mixed into a tank containing water until the granules or crystals are dissolved and the desired concentration is reached. The solution is then injected into the irrigation system.

The injection point should be located so that the injected fertilizer and the irrigation water can become thoroughly mixed well upstream of any branching of the flow. Injecting at a point in front of the filter will reduce the chance of contaminants entering the system. The system should be allowed to fill and come up to full pressure before injection begins. Following injection, the system should be operated to flush the fertilizer from the lines. Leaving residual fertilizer in the line may encourage clogging from chemical precipitates or organic sources such as bacterial slimes.

Nitrogen

The fertilizer most commonly injected is nitrogen, with many soluble nitrogen sources working well in fertigation.

Following is a list of common nitrogen sources with comments about their use in fertigation:

Anhydrous ammonia or aqua ammonia. These nitrogen sources cause an increase in water pH, which may result in a precipitate if calcium or magnesium are present in the irrigation water along with comparable levels of bicarbonate. Volatilization of nitrogen (loss to the atmosphere) may also occur when anhydrous or aqua ammonia are used.

Urea. Urea is relatively soluble in the irrigation water and is not strongly held by soil particles, so it moves deeper into the soil than ammonia products. Urea is transformed by hydrolysis into ammonium, which is then fixed to the soil particles.

Ammonium sulfate, Ammonium nitrate, and Potassium nitrate. All are relatively soluble in water and cause only a slight shift in the soil or water pH.

Calcium nitrate. Calcium nitrate is relatively soluble in water and causes only a slight shift in the soil or water pH. If the water is high in bicarbonate, however, the calcium content may lead to precipitation of calcium carbonate (lime).

Ammonium phosphate. Ammonium phosphate can also cause soil acidification. If calcium or magnesium levels are high enough in the irrigation water, precipitates may also form, which can clog the drip emitters. (See the discussion under ***Phosphorus***, below, for precautions in using ammonium phosphate.)

Phosphorus

Using phosphate fertilizers may cause chemical or physical precipitate clogging. The calcium and magnesium content and the pH of the irrigation water should be considered, since calcium phosphate and magnesium phosphate precipitates may form when the water pH is higher than 7.5. Acidifying the water with sulfuric acid or using phosphoric acid keeps the irrigation water pH low and minimizes precipitation problems.

Phosphorous is quickly fixed to soil particles and does not move readily into the soil profile, but it has been found to move much more easily under drip irrigation than under conventional irrigation methods.

Potassium

Injecting potassium fertilizers usually causes few problems, but caution should be observed if potassium fertilizers are mixed with other fertilizers. Potassium, like phosphorous, is fixed by clay particles and does not move readily through the soil profile.

Potassium is usually applied in the form of potassium chloride, but for crops sensitive to chloride, potassium sulfate or potassium nitrate may be preferable. Potassium sulfate is not very soluble and may not dissolve well in the irrigation water.

Fertilizers for Drip Irrigation

By Blaine Hanson, UC Irrigation and Drainage Specialist

Units of Concentration

Fertilizer concentrations normally are expressed in parts per million or percent concentration. Both units are on a weight basis.

One ppm means one pound of fertilizer material dissolved in one million pounds of water or solution, while percent concentration is the ratio of the weight of the fertilizer material in the solution to the weight of the solution. The ratio is multiplied by 100 to obtain the percent concentration.

Other units of concentration are milligrams per liter, milligrams per kilograms, ounces by weight per gallon, and pounds per gallon. Some conversion factors are:

1 ppm = 1 milligram per liter
1 ppm = 1 milligram per kilogram
1 percent = 10,000 ppm
1 percent = 1.33 ounce by weight per gallon of water
0.1 percent = 1000 ppm

Fertilizer Characteristics

Fertilizer characteristics include their nutrient content as indicated by the grade designations, solubility, density, salting out temperature, salt index, and acidity.

Fertilizer Designation. Fertilizers are designated by their primary nutrient contents. These contents are nitrogen (N), phosphorus pentoxide (P_2O_5), and potash (K_2O). The order of designation is N - P_2O_5 - K_2O, normally listed as N - P - K. The number is the grade of the fertilizer and is the weight percent of the nutrient contained in the fertilizer.

It may be desirable to convert the % phosphorus pentoxide to % phosphorus (% P) and the % potash to % potassium (% K). The following conversion factors can be used:

$$\% P \times 2.29 = \% P_2O_5$$
$$\% P_2O_5 \times 0.44 = \% P$$
$$\% K \times 1.20 = \% K_2O$$
$$\% K_2O \times 0.83 = \% K$$

Solubility. The solubility of a fertilizer is the amount of dry chemical that can be dissolved in a given amount of water. Solubility increases as the temperature of the water increases as shown in Figure 1. Some companies recommend mixing the fertilizer in hot water to better dissolve the material. Examples of solubilities of dry fertilizers are in Table 5.

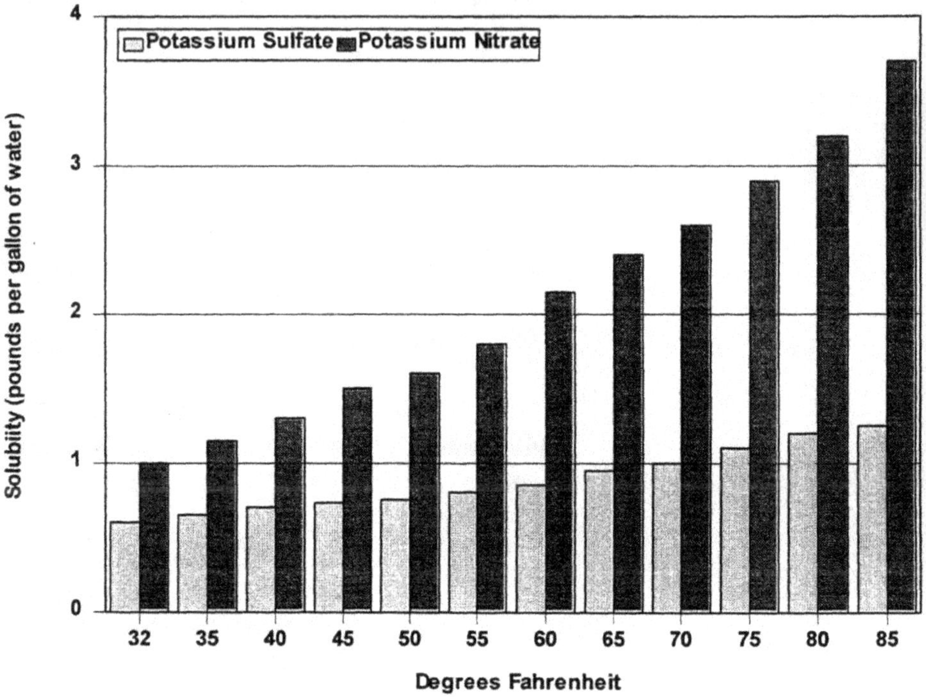

Figure 1. Solubility with temperature.

Density. The density of a fertilizer solution is the ratio of the weight of the solution of a known volume to the volume of the solution. Units of density are normally in pounds per gallon of solution. The density multiplied by the percent content of a nutrient divided by 100 gives the pounds of the nutrient in the solution. Densities of some liquid fertilizers are listed herein.

Salting Out Temperature. This is the temperature at which the fertilizer salt precipitates out of the solution. A fertilizer should not be used where air temperatures are less than its salting out temperature.

Salt Index. The salt index of a fertilizer is a measure of its contribution to the salinity of the soil solution. The lower the salt index, the less the contribution. Table 1 lists salt indices for some fertilizer using the salt index of sodium nitrate as the standard (salt index = 100). A salt index of 116, that of potassium chloride, contains 16 percent more salt that does sodium nitrate.

Effect on soil pH. Acid fertilizers tend to reduce the soil pH, while basic fertilizers tend to increase the soil pH. Neutral materials have no effect on soil pH. Table 4 lists qualitative effects on pH for some fertilizers.

Table 1. Relative salt index for some fertilizers.

Fertilizer	Salt Index	Partial
Sodium nitrate (nitrate of soda)	100	6.1
Potassium chloride	116	1.9
Ammonium nitrate	102	3.0
Urea	73	1.6
Potassium nitrate	74	1.3
Ammonium sulfate	69	3.3
Calcium nitrate	53	3.4
Potassium sulfate	46	0.9
Gypsum	8	-
Ammonium thiosulfate	84	-
Urea-ammonium nitrate solution (UAN-32)	95	-
Potassium thiosulfate	64	-

*Salt Index/20 lb of nutrient

Considerations and Constraints

Fertilizer irrigated into drip irrigation systems must be soluble. It must remain in solution when injected into the irrigation water and not cause any precipitate that can clog emitters. Other factors to be considered are the cost of the material, storage requirements, and ease of handling.

Questions have arisen concerning the maximum nitrogen concentration in the irrigation water that causes root damage. Little documented information appears to exist on this matter. However, one source suggests a maximum concentration of 5 percent (50,000 ppm) with acceptable levels of 1 to 2 percent (10,000 to 20,000 ppm).

Nitrogen

Forms of nitrogen in nitrogen fertilizers are urea, ammonium, and nitrate. Urea is a neutral molecule (no electric change) formed by reacting ammonia with carbon dioxide. It is highly soluble in water. However, when incorporated into the soil, soil enzymes cause urea to hydrolyze into the ammonium ion and carbon dioxide.

Ammonium nitrogen is a positively charged ion. When incorporated into the soil, the ammonium ion quickly becomes attached to the soil particles and as a result does not move readily with soil water movement. Absorption of ammonium ion is greater in clay soils and soils with high organic matter content than in sandy soils. The ammonium ion is converted to nitrate by a process called nitrification. The rate of nitrification depends on the soil temperature and soil water content.

Nitrate is a negatively charged ion. Thus, it readily moves in the soil with water because soil particles have a negative charge which repels nitrate ions. Nitrate can be lost from the soil by a process called denitrification. This process, which occurs in soils with excessive soil moisture, results when anaerobic bacteria use the oxygen in the nitrate ion. This forms gaseous compounds of nitrogen which are released from the soil.

Nitrogen Fertilizers for Drip Irrigation

Ammonium nitrate (AN-20). This liquid fertilizer is formed by reacting ammonia with nitric acid. It contains 20% nitrogen (20-0-0). The density of this fertilizer is 10.5 pounds per gallon, each gallon containing 2.1 pounds of N. The salting out temperature is 41°F. This fertilizer should never be mixed with acids or acid solutions. The percent concentration of nitrate N is 10% and that of ammonia N is 10%.

Urea-ammonium nitrate (UAN-32). This fertilizer is formed by reacting urea and ammonia with nitric acid. Its nitrogen content is 32% (32-0-0). The forms and concentrations of N are urea N (16.4%), ammonia N (7.4%), and nitrate N (7.4%). The density of UAN-32 is 11.1 pounds per gallon, each gallon containing 3.5 pounds of N. The salting out temperature is 32°F.

In colder climates, a urea-ammonium nitrate solution containing 28% N is used. The salting out temperature of this fertilizer is 1°F. The density of this solution is 10.7 pounds per gallon, each gallon containing nearly 3 pounds of N.

Calcium ammonium nitrate (CAN-17). This fertilizer, designated as 17-0-0, is formed by reacting ammonia and calcium carbonate (lime) with nitric acid. The nitrogen content is 17% (11.2% nitrate N and 5.8% ammonia N). Its calcium content is 7.6% to 8.8%. The density of CAN-17 is 12.6 pounds per gallon. Its salting out temperature is 30°F.

Calcium nitrate (CN-9). This fertilizer, designated as 9-0-0, contains 9% nitrogen consisting of 8.42% nitrate nitrogen and 0.58% ammonium nitrogen. It also contains 11% calcium. Density of the fertilizer is 12.2 pounds per gallon, and its salting out temperature is 0°F-5°F. Its pH ranges from 5.0 to 5.4.

Aqua ammonia is formed by dissolving ammonia in water. Its nitrogen content is 20% (20-0-0). Use of this fertilizer can increase the pH of the irrigation water, thus causing potential problems with carbonate precipitation.

Urea-sulfuric acid. This fertilizer is a combination of urea and sulfuric acid commonly known as N-pHURIC made by UNOCAL although another company markets a similar material under a US tradename. Table 1 lists the various solutions of N-pHURIC.

Urea-sulfuric acid can be mixed with other fertilizers such as phosphoric acid, ammonium polyphosphate (10-34-0), potassium sulfate, and potassium chloride. In some cases, however, the order at which the materials are added is important. Guidelines are provided by UNOCAL. Under no circumstances should any fertilizer containing nitrate-nitrogen be mixed with urea-sulfuric acid.

Table 2. Various solutions of N-pHURIC.

%N	% Sulfuric Acid	Density (lbs per gallon)	Salting Out Temperature (°F)
10	55	12.8	34
15	49	12.7	22
28	27	11.8	39

Phosphorus

Several forms of phosphorus fertilizers exist. However, use of these materials in irrigation water with calcium concentrations exceeding about 40 to 50 ppm (2 to 2.5 milliequivalent per liter) can result in the precipitation of calcium phosphate. This precipitate can clog a drip irrigation system.

Ammonium polyphosphates are made by combining superphosphoric acid with ammonia. A variety of grades are listed in Table 3.

Table 3. Grades of ammonium polyphosphate.

% N	% P_2O_5	Density (lb/gal)
8	24	10.5
9	30	11.3
10	34	11.4
11	37	11.7

Phosphoric acid comes in two forms: white (0-54-0) and green (0-52-0). White phosphoric acid is a more pure acid than green acid but it is also more expensive. Normally, the green phosphoric acid is used for drip irrigation. The density of phosphoric acid is about 14 pounds per gallon.

Phosphoric acid must be injected at a rate sufficient to lower the pH of the irrigation water enough to prevent phosphate precipitation. A pH of about 5 or less is recommended.

Potassium

Common solutions for injecting potassium into drip irrigation systems are in Table 4.

Table 4. Liquid potassium fertilizers.

Fertilizer	Designation	Density (pounds per gallon)	Solubility (pounds per gallon of water)	Salting Out Temperature (°F)	Other
Potassium chloride (granular)	0-0-60	-	2.9 (70°F)	-	-
Potassium nitrate (granular)	13-0-45	-	2.8 (70°F)	-	-
Potassium nitrate (liquid)	3-0-11 2-0-7	9.8 9.4	- -	60 -	- -
CAN-17 with potassium nitrate (liquid)	17-0-4	11.6	-	30	4% calcium
Urea-Potassium nitrate (liquid)	15-0-5	9.76	-	25	-
Potassium sulfate (granular)	0-0-50	-	1.0 (70°F)	-	-
Potassium sulfate (ESP) (liquid)	1-0-8	10.3	-	45	2.5% sulfur
Potassium thiosulfate (liquid)	0-0-25	12.2	-	low	17% sulfur

Use of potassium chloride may cause chloride toxicity problems with trees and vines. Its solubility, however, is higher than those of potassium sulfate and potassium nitrate. Potassium sulfate has the lowest solubility.

Gypsum

Gypsum is sometimes injected into drip systems to improving water infiltration and to change calcium-magnesium ratios although information on the latter is sketchy. Because the solubility of gypsum in water is very low, injection machines are used which inject a slurry of dissolved and suspended gypsum particles into the irrigation water. Caution should be used in injecting gypsum into irrigation water containing more than 2 milliequivalent per liter of bicarbonate because gypsum injection could cause calcium carbonate precipitation. (The solubility of calcium carbonate is 0.000051 pounds per gallon.)

Solid Fertilizers

A variety of solid or dry fertilizers are available. Use of these fertilizers requires mixing them in water. The amount of fertilizer that can be dissolved in water depends on its solubility. Generally, dry nitrogen fertilizers have a high solubility while dry potassium fertilizers have a lower solubility. The higher the temperature the higher the solubility. Table 4 lists some dry fertilizers and their solubilities. A gypsum injector machine may be necessary to inject sufficient material for fertilizers with low solubility.

Table 5. Dry fertilizers (all solubilities are for 32°F for comparison purposes).

Fertilizer	Designation	Solubility (pounds per gallon of water)	Effect on Soil pH
Ammonium nitrate	33.5-0-0	10.0	A
Ammonium sulfate	20-0-0	5.9	A
Calcium nitrate	15.5-0-0	8.5	B
Sodium nitrate	16-0-0	6.1	B
Potassium nitrate	13-0-62	1.1	N
Potassium chloride	0-0-53	2.9	N
Potassium sulfate	0-0-53	0.6	N
Monoammonium phosphate	11-48-0	1.9	A
Diammonium phosphate	21-53-0	3.6	A
Urea	45-0-0	6.5	A

A = acid; B = basic; N = neutral.

Organic Fertilizers

A limited number of organic fertilizers are available for injecting into drip irrigation systems. Suspensions of fish meal or blood meal have been used, but some clogging problems have been reported. A water-soluble nitrogen fertilizer, Phytamin 800 (7-0-0) liquid solution, is available for injection. The density of this material is about 10 pounds per gallon.

Soda of nitrate (16-0-0) or sodium nitrate (commonly called Chilean nitrate) can be used for about 20 percent of the crop's nitrogen requirement. Higher applications may cause water infiltration problems caused by excessive levels of sodium in the soil. The solubility of sodium nitrate is about 7 pounds per gallon of water. Its density is about 15.3 pounds per gallon of solution.

Blended Fertilizers

The fertilizers listed herein are some of the more commonly used materials. However, there are many blends of these materials to form multinutrient fertilizers. These are too numerous to list. Growers using these fertilizers must obtain the technical information from the supplier or manufacturer for technical information.

The advantage of blended fertilizers is that more than one type of nutrient can be applied at a time thus saving time and labor. The disadvantage of these fertilizers is that an application rate based on the needs for one type of nutrient may result in an excessive application of another nutrient.

Micronutrients

Micronutrients such as manganese, zinc, copper, iron, molybdenum, and boron can be applied with drip irrigation. Fertilizers commonly used to apply manganese, zinc, copper, and iron are sulfate, citrate, and/or chelate solutions. Molybdenum and boron are frequently applied as sodium molybdate and sodium borate.

Costs

Fertilizer costs should be considered in selecting a fertilizer. Costs comparisons of fertilizer containing the same nutrient should include not only the unit costs, i.e., cost per ton, but also the cost per pound of nutrient. Fertilizers with about the same unit costs expressed in dollars per ton may have very different costs based on dollars per pound of nutrient. The nomograph in Appendix E can be used for determing the costs on a nutrient basis.

References

Chemigation. 1996. The Irrigation Association.

Ricardo Silberman (Personal Communication). Chilean Nitrate Corporation, 5528 N. Palm Ave, Suite 115, Fresno, CA 93704.

Tessende-Kerley, Phoenix, AZ.

UNOCAL. 1996. N-pHURIC Reference Manual

Les Breschini (Personal Communication). John Pryor, Co. Salinas, CA.

Western Fertilizer Handbook. 1995. California Fertilizer Association.

Applying the Right Amount of Fertilizer

By Blaine Hanson, UC Irrigation and Drainage Specialist

Drip irrigation makes it possible to match the amount of nitrogen fertilizer or other fertilizers to the needs of the plant at each growth stage, thereby improving crop yield, reducing fertilizer costs, and cutting down on nitrate leaching.

Applying the right amount of fertilizer requires knowing:

- the fertilizer requirements of the crop at each growth stage
- fertilizer concentrations in the irrigation water
- the fertilizer level of the soil

Table 1 gives the nitrogen requirements of common vegetable crops at various growth stages. Table 2 gives seasonal requirements of nitrogen, phosphorus, and potassium for some crops. As these tables illustrate, nitrogen needs vary widely from crop to crop and from one growth stage to another.

The nitrogen status of a crop can be monitored by analyzing plant tissue. As a sampling procedure, T.K. Hartz of the Department of Vegetable Crops at the University of California, Davis recommends collecting at least 30 petioles or midribs, selecting the youngest, fully expanded leaves (usually situated three or four leaves back from the growing point), placing the samples in plastic bags, and storing them on ice in a cooler until they can be processed.

Table 3 gives petiole nitrate sufficiency values for drip-irrigated vegetable crops.

Irrigation Water Nitrogen

To determine how much nitrogen (N) should be applied, subtract the nitrogen concentrations of the irrigation water and soil from the nitrogen requirement of the crop.

The following equations convert nitrate concentrations (expressed in parts per million [ppm]) to pounds of nitrogen per acre-inch of water. Use *Equation 1* if the water analysis shows ppm of nitrogen. Use *Equation 2* if the analysis shows only the nitrate concentration.

$$\text{1 ppm of nitrogen} = \text{0.23lbs. of N per acre-inch of irrigation water} \quad (1)$$

1 ppm of nitrate (NO^{3-}) = 0.051 lbs. of N per acre-inch of irrigation water (2)

Multiplying the pounds of N per acre-inch by the inches of applied water gives the number of pounds of nitrogen per acre in the irrigation water.

Example

How much nitrogen should be applied where the nitrogen concentration of the irrigation water is 5 ppm and the total amount of applied water is three inches?

pounds of N per acre-inch = 5 x 0.23 = 1.1

total lbs. to be applied per acre = 1.1 lbs. per acre-inch x 3 inches
= 3.3 lbs. of N per acre

If the crop requirement is 10 pounds per week, then about 7 pounds of N fertilizer should be applied.

Calculating the Injection Rate

The injection rate can be calculated with the following equation:

$$IR = \frac{P_N \times A \times 100}{P_G \times \% N \times IT}$$

where IR = the injection rate in gallon per hours
P_N = pounds of fertilizer per acre
A = acres irrigated
P_G = density of fertilizer solution in pounds per gallon (Table 4)
% = percent nutrient in the fertilizer (Table 4)
IT = injection time in hours

A step-by-step procedure for calculating the injection time is:

1. Determine the pounds of fertilizer per acre per week to be applied using the data in Table 1 for nitrogen during each irrigation.

2. Determine the pounds of N per gallon of fertilizer using Table 4 for nitrogen.

3. Determine the gallons of fertilizer needed per acre per week for each irrigation by:

gallons/acre = pounds per acre/pounds per gallon

4. Determine the total gallons needed per week:

gallons = gallons per acre x acres

5. Determine the injection rate given the desired injection time during the irrigation set.

Injection rate (gallons/hour) = gallons/injection time.

Example

Calculate the injection rate needed to apply 15 pounds per acre per week of fertilizer to a 40 acre drip irrigated tomato field. The field is irrigated twice per week in two sets per irrigation, 20 acres per set. The irrigation set time is 12 hours. The fertilizer is UAN-32. 7.5 pounds of N per acre are to be applied during each irrigation. The injection time per irrigation set is 10 hours.

1. 15 pounds of N per acre per week are needed or 7.5 pounds per irrigation.
2. From Table 4, one gallon of UAN-32 contains 3.5 pounds of N.
3. The gallons of fertilizer needed per acre are:

gallons/acre = 7.5 pounds of N per acre/3.5 pounds of N per gallon
= 2.1 gallons

4. Total gallons needed for each irrigation set are 2.1 gallons per acre x 20 acres = 42 gallons.
5. The injection rate is 42 gallons/10 hours = 4.2 gallons per hour.

Concentrations

The concentrations of the fertilizer in irrigation can be determined using the following equations:

$$C_I = \frac{C_F \times IR}{60 \times Q}$$

where C_I = fertilizer concentration in irrigation water (ppm)
C_F = fertilizer concentration in liquid solution (1% = 10,000 ppm)
IR = injection rate (gallons per hour)
Q = irrigation system flowrate (gallons per minute)

Example

Calculate the nitrogen concentration in the irrigation water where the system flowrate is 300 gpm, the injection rate is 4.2 gph, and the fertilizer is UAN-32 (C_F - 32% x 10,000 ppm = 320,000 ppm).

$$C_I = \frac{320{,}000 \text{ ppm} \times 4.2 \text{ gph}}{60 \times 300 \text{ gpm}} = 75 \text{ ppm}$$

Injection Time

How long should the injection time be compared with the irrigation set time. One approach used is the 1:2:1 method. For the first quarter of the irritation time, no injection occurs thus allowing the pipelines and laterals to fill and preventing fertilizer from moving to the outer part of the wetted pattern. During the middle half of the irrigation, injection occurs. No injection occurs during the last quarter of the irrigation which allows the pipelines and laterals to be flushed of the fertilizer and to move the fertilizer in the soil away from the emitter. For example, if the irrigation set time is four hours, no injection occurs during the first and last hours. Injection only occurs during the middle two hours.

Reference

Hartz, T.K. 1993. *Drip irrigation and fertigation management of vegetable crops.* Department of Vegetable Crops, University of California, Davis.

Table 1. Nitrogen requirements of vegetable crops.

crop	*growth stage*	*approximate nitrogen requirement (pounds/acre/week)*
broccoli	*early growth*	*5 - 15* [1]
	mid-season	*10 - 20*
	button formation	*15 - 30*
	head development	*10 - 20*
cucumber	*vegetative growth*	*5 - 10*
	early flowering/fruit set	*10 - 20*
	fruit bulking	*10 - 15*
	first harvest	*5 - 10*
lettuce	*early growth*	*5 - 10*
	cupping	*10 - 20*
	head filling	*15 - 30*
melon	*vegetative growth*	*5 - 10*
	early flowering/fruit set	*10 - 20*
	fruit bulking	*10 - 15*
	first harvest	*5 - 10*
pepper	*vegetative growth*	*5 - 10*
	early flowering/fruit set	*15 - 30*
	fruit bulking	*15 - 20*
	first harvest	*5 - 10*
squash	*vegetative growth*	*5 - 10*
	early flowering	*10 - 20*
	first harvest	*5 - 10*
tomato	*vegetative growth*	*5 - 10*
	early flowering/fruit set	*15 - 20*
	fruit bulking	*10 - 15*
	first harvest	*5 - 10*

[1] *Higher values represent fertilization needs in soils with small amounts of residual nitrogen or where high temperatures cause rapid plant growth.*

Source: Hartz, T.K. 1993. *Drip irrigation and fertigation management of vegetable crops.* Department of Vegetable Crops, University of California, Davis.

Table 2. Seasonal nutrient requirements.

Crop	*Yield*	*Pounds per Acre* N	P_2O_5	K_2O
Field crops				
Corn (grain)	*5 t. (179 bu.)*	*240*	*100*	*240*
Cotton (lint)	*1,500 lbs.*	*180*	*65*	*125*
Oats	*3,200 lbs. (100 bu.)*	*115*	*40*	*145*
Safflower	*4,000 lbs.*	*200*	*50*	*150*
Sugar Beets	*30 t.*	*255*	*60*	*550*
Vegetable crops				
Asparagus	*3,000 lbs.*	*95*	*50*	*120*
Beans (snap)	*10,000 lbs.*	*175*	*40*	*200*
Broccoli	*18,000 lbs.*	*80*	*30*	*75*
Cabbage	*35 t.*	*270*	*65*	*250*
Celery	*75 t.*	*280*	*165*	*750*
Lettuce	*20 t.*	*95*	*30*	*200*
Potatoes (Irish)	*500 cwt.*	*270*	*100*	*550*
Squash	*10 t.*	*85*	*20*	*120*
Sweet potatoes	*15 t.*	*155*	*70*	*315*
Tomatoes	*30 t.*	*180*	*50*	*340*
Fruit and nut crops				
Almonds (in shell)	*3,000 lbs*	*200*	*75*	*250*
Apples	*15 t.*	*120*	*55*	*215*
Cantaloupes	*30 t.*	*220*	*70*	*400*
Grapes	*15 t.*	*125*	*45*	*195*
Oranges	*30 t.*	*265*	*55*	*330*
Peaches	*15 t.*	*95*	*40*	*120*
Pears	*15 t.*	*85*	*25*	*95*
Prunes	*15 t*	*90*	*30*	*130*

Source: Western Fertilizer Handbook. 1995. California Fertilizer Association.

Table 3. Petiole NO_3-N sufficiency values for drip-irrigated vegetables.

crop	*growth stage*	*Petiole NO_3 - N concentration (ppm)* dry tissue	fresh sap
broccoli	*mid -growth*	*10,000 - 20,000*	*1200 - 2400*
	button formation	*8000 - 15,000*	*1000 - 1600*
	preharvest	*5000 - 8000*	*800 - 1200*
cantaloupe	*early flower*	*12,000 - 15,000*	*1000 - 1200*
	fruit bulking	*8000 - 10,000*	*800 - 1000*
	first harvest	*4000 - 6000*	*700 - 800*
celery	*mid-growth*	*7000 - 10,000*	*300 - 500*
	preharvest	*6000 - 8000*	*300 - 400*
lettuce	*early head formation*	*7000 - 12,000*	*400 - 600*
	preharvest	*6000 - 10,000*	*300 - 500*
pepper	*vegetative growth*	*7000 - 10,000*	*900 - 1200*
	early flower/fruit	*5000 - 8000*	*700 - 1000*
	fruit bulking	*5000 - 8000*	*700 - 1000*
	first harvest	*5000 - 7000*	*700 - 900*
tomato	*vegetative growth*	*10,000 - 14,000*	*700 - 900*
	early flower/fruit	*9000 - 12,000*	*600 - 800*
	fruit bulking	*6000 - 8000*	*500 - 700*
	preharvest	*4000 - 6000*	*400 - 600*
watermelon	*early flowering*	*12,000 - 15,000*	*900 - 1100*
	fruit bulking	*8000 - 15,000*	*700 - 900*
	fruit harvest	*5000 - 8000*	*500 - 700*

Source: Hartz, T.K. 1993. *Drip irrigation and fertigation management of vegetable crops.* Department of Vegetable Crops, University of California, Davis.

Table 4. Liquid weights and nitrogen amounts for various fertilizers.

fertilizer	*lbs. per gallon*	*% nitrogen*	*lbs. N per gallon*
ammonium nitrate (AN 20)	*10.7*	*21*	*2.2*
ammonium thiosulfate	*11.0*	*12*	*1.3*
aqua ammonia	*7.5*	*20*	*1.5*
urea ammonium nitrate (UAN 32)	*11.1*	*32*	*3.5*
N-pHURIC 28/27	*11.84*	*28*	*3.3*
N-pHURIC 15/49	*12.65*	*15*	*1.9*
N-pHURIC 10/55	*12.80*	*10*	*1.3*
CAN 17	*12.65*	*17*	*2.1*
CN-9	*12.2*	*9*	*1.1*
US 28	*11.87*	*28*	*3.3*
US 15	*12.65*	*15*	*1.9*
US 10	*12.80*	*10*	*1.3*

Fertilizer Patterns Under Drip Irrigation

By Blaine Hanson, UC Irrigation and Drainage Specialist

When fertilizer is applied to a field being irrigated with a drip system, it becomes distributed through the soil according to the wetting patterns, but depending also on the method used to apply the fertilizer, the kind of fertilizer used, the soil type, the soil profile, and how evenly the water is applied over the field.

To benefit the crop, the fertilizer must be placed into that portion of the soil being wetted by the emitter. Water flowing from the emitter then distributes the fertilizer throughout the root zone, which generally mirrors the shape of the wetted soil.

Fertilizer can be applied by broadcasting, by banding, or by injecting into the drip system. Broadcasting is generally ineffective under drip irrigation simply because so little of the fertilizer ends up in the soil wetted by the emitter. Banding—applying the fertilizer in bands across the soil surface—can be more effective if the fertilizer band is carefully placed in relation to the emitter and the plant, ideally between the emitter and the plant so that water flowing from the emitter will carry the dissolved fertilizer toward the roots. The band should be as close to the emitter as possible so that the fertilizer is well distributed throughout the root zone.

If the fertilizer is injected into the drip system, the distribution pattern will be determined by the patterns of water flow around the drip tape, by how readily the kind of fertilizer used moves with the water, by the level of fertilizer concentrations in the water, and by how the injection is timed in relation to irrigation. Some fertilizers are highly mobile, while others tend to become adsorbed to the soil and do not move with the water.

Nitrogen

Nitrate is a highly mobile ion that moves readily with water. Nitrate patterns are therefore strongly affected by water flow patterns in the soil Frequent nitrate fertilizer injections will distribute the fertilizer throughout the wetted soil, with the highest nitrate concentrations settling near the emitter. Figure 1 illustrates this pattern. If injections are infrequent compared to irrigations, the nitrate will tend to be leached out of the soil near the emitter and accumulate near the edge of the wetted pattern. Under a subsurface drip system, nitrate can also accumulate near the surface at depths shallower than the drip tape.

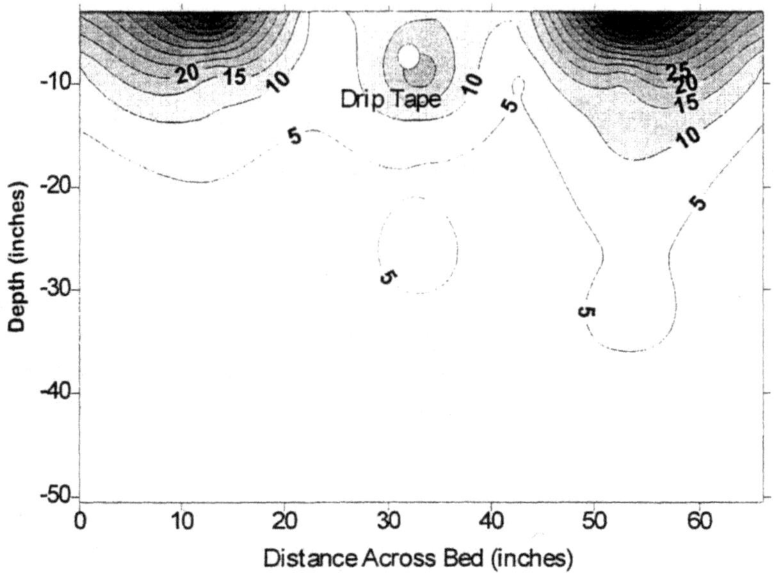

Figure 1. Nitrate patterns resulting from injecting nitrogen fertilizer through a drip irrigation system.
(Nitrate values are in ppm.)

Injecting small amounts of ammonium fertilizer will cause ammonium ions to be adsorbed to the soil in the vicinity of the emitter, while injecting large enough amounts for the adsorption capacity of the soil to become satisfied will cause the ammonium ions to move deeper into the soil. How deeply the ions penetrate will depend on the cation exchange capacity of the soil and on how much fertilizer is applied. In the space of two or three weeks the ammonium will be converted to nitrate.

Since urea is relatively soluble in water and is not strongly adsorbed by the soil, it moves more readily with water in the soil than ammonium. The urea will convert to ammonium, which will be adsorbed to the soil.

Phosphorus

Phosphorus tends to be adsorbed by the soil, and, depending on soil type, may not move readily with the water. This is a particular problem in clay soils or in soils with a high cation exchange capacity, but less of a problem in sandy soils. Because of its adsorption tendency, if phosphorus is broadcast onto the soil surface, it may move into the soil only a few inches. Banding the fertilizer along the top of the root zone may yield better results, but research has found that injecting phosphorus through a drip system can more effectively promote its distribution through the root zone.

How much fertilizer is applied in a given period also appears to affect the distribution pattern. The more fertilizer applied, the better the distribution. One study showed that phosphorus applied at the rate of 6 pounds per acre in a clay loam soil moved about 5 inches from the emitter, while at 35 pounds per acre it moved nearly 11 inches.

Soil type also has an effect. One study showed that phosphorus moved twice as far horizontally and three times as far vertically in sandy soil than it did in clay soil. Yet, if the soil has a very high adsorption capacity, phosphorus may move only a short distance from the emitter. A study illustrating this point found that phosphorus applied at 132,535, and 1070 pounds per acre moved only a few inches from a buried drip emitter. Some experts believe, however, that even this limited phosphorus movement is sufficient to meet the plant's phosphorus requirement.

Potassium

Potassium fertilizers such as potassium sulfate, potassium chloride, and potassium nitrate are quite water-soluble. Like ammonium, potassium will be adsorbed to soil particles. Increasing the potassium concentration in the irrigation will cause it to move more deeply into the soil.

Unfortunately, not much is known about the movement in the soil of other fertilizers under drip irrigation.

Field-wide Uniformity

How fertilizer injected into a drip system becomes distributed over the field is determined by the extent to which the emitters or tape discharge at the same rate. The more uniform the discharge rates, the more evenly the fertilizer will be applied. Figure 2 below shows how variation in emitter discharge rates correlates with how uniformly nitrogen is applied along a drip lateral buried at six inches below the surface.

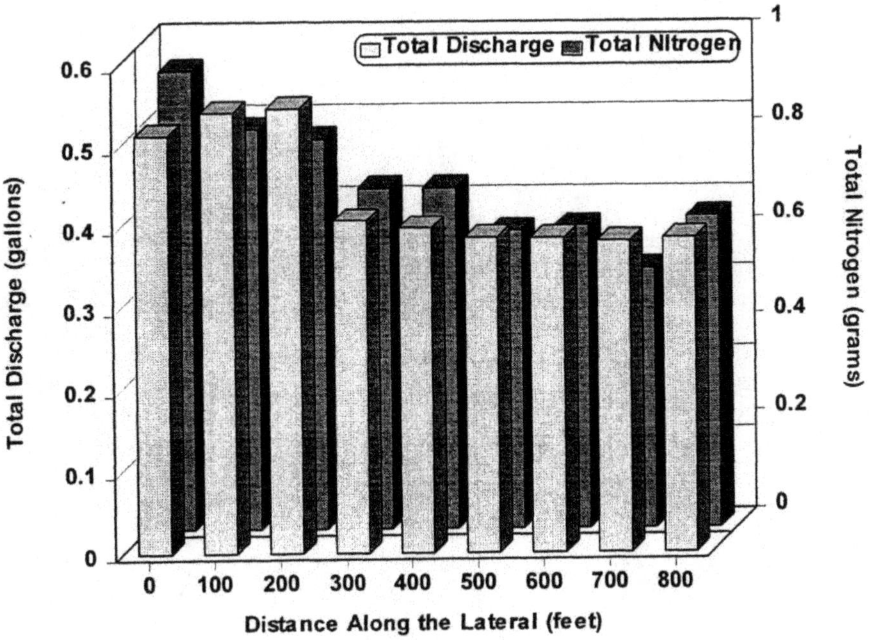

Figure 2. The effect of emitter discharge rate variation on fertilizer application uniformity.

References

Bar-Yosef, B., and M. R. Scheikholslami. 1976. "Distribution of water and ions in soils irrigated and fertilized from a trickle source." *Proceedings of the Soil Science Society of America.* Vol. 40: 575-82.

Chase, R. G. 1985. "Phosphorus application through a subsurface trickle system." In: *Drip/Trickle Irrigation in Action, Proceedings of the Third International Drip/Trickle Irrigation Congress,* Fresno, California.

Haynes, R. J. 1985. "Principles of fertilizer use for trickle irrigated crops." Fertilizer Research, Vol. 6: 235-55.

Mikkelsen, R. L. 1989. "Phosphorus fertilization through drip irrigation." *Journal of Production Agriculture,* Vol. 2: 279-86.

Miller, R. J., D. E. Rolston, R. S. Rauschkolb, and D. W. Wolfe. 1976. "Drip application of nitrogen is efficient." *California Agriculture,* Vol. 30: 16-18.

Rolston, D. E., R. S. Rauschkolb, C. J. Phene, R. J. Miller, K. Uriu, R. M. Carlson, and D. W. Henderson. 1979. *Applying nutrients and other chemicals to trickle-irrigated crops.* University of California Bulletin No. 1893.

Water-Fertilizer Interactions

By Blaine Hanson, UC Irrigation and Drainage Specialist

Although injecting fertilizer through a drip system offers a number of important benefits, irrigators should also be aware that fertilizer can interact with the irrigation water to form precipitates that may clog the drip system.

Anhydrous and Aqua Ammonia

Injecting anhydrous or aqua ammonia increases the pH of the irrigation water. If the pH increases to more than 7.5, and if the irrigation water contains at least 2 meq/liter of bicarbonate, calcium carbonate can precipitate.

Calcium

If the irrigation water has more than 2 meq/liter of bicarbonate and the pH of the water is more than 7.5, adding fertilizer containing calcium — such as calcium nitrate — may cause calcium carbonate to precipitate.

Sulfate

If fertilizers containing sulfate, such as potassium sulfate, are added to irrigation water having a calcium concentration of more than 20-30 meq/liter, gypsum may precipitate.

Phosphorus

Many phosphorous fertilizers are not readily soluble in water and may therefore react with calcium and magnesium in the irrigation water to form phosphate precipitates if the calcium or magnesium concentrations are higher than 2.5 meq/liter or 40-50 ppm. These precipitates can be difficult to dissolve out of emitters.

To prevent phosphate precipitates from forming, the irrigation water can be acidified to lower the pH to less of 5 or less, although this low pH may damage some kinds of emitters. Injecting a slug of highly concentrated phosphoric acid is sometimes effective, but the amount of acidic phosphorous fertilizer needed to reduce the water pH may greatly exceed the amount needed for fertilizer.

Mixing Fertilizers Containing Calcium and Sulfate

Mixing a fertilizer containing calcium with a fertilizer containing sulfate can cause gypsum to precipitate. One example of this would be mixing calcium nitrate with potassium sulfate. While both of these fertilizers are water-soluble, mixing them together into irrigation water will cause calcium sulfate or gypsum to form, which is much less soluble and which will precipitate out of the water.

Testing the Irrigation Water

Before a drip system is used to inject fertilizer, the irrigation water should be tested for any potential to form precipitates through water-fertilizer interaction. As a first step, the water should be analyzed for calcium, magnesium, bicarbonate, and sulfate content. Concentrations of any of these elements exceeding 2.5 meq/liter suggest a potential to form precipitates when mixed with certain fertilizers. Next the fertilizer should be mixed into a container of irrigation water at the concentrations desired for fertigation. The mixture should then be covered and held for the length of time approximating the period the fertilizer would be present in the irrigation system. If the water turns cloudy or white, or if a precipitate collects on the bottom of the container, precipitation will probably result from injecting the fertilizer into the irrigation water. This testing procedure may have to be repeated frequently, particularly when phosphate fertilizers are used, since characteristics may differ by manufacturer.

Guidelines for Mixing Fertilizer

Figure 1 can be used as a guideline for determining if mixing fertilizers with water or other chemicals may create a potential clogging problem. This figure was developed by UNOCAL. In general, avoid mixing fertilizers unless clear guidelines are available.

Example

Assess any potential compatibility problems caused by injecting calcium nitrate into water.

From Figure 1, the product number for water is 28 and the product number for calcium nitrate is 33. At the intersection of how 28 and column 33, a light gray square is found indicating a compatibility problem with certain limits. Injecting calcium nitrate into a water containing more than 2 meg/l of bicarbonate could cause calcium carbonate precipitation. The calcium nitrate would provide a calcium source.

1 **Anhydrous Ammonia**
2 **Aqua Ammonia; 20-0-0**
3 **Urea solution; 23-0-0**
4 **Ammonium nitrate soln; AN20**
5 **Urea-ammonium nitrate soln; UAN32**
6 Urea-ammonia soln; 33-0-0
7 **DI-AN soln; 18-0-0**
8 Ammonium sulfate soln
9 Ammonium phosphate soln; 8-24-0
10 Ammonium polyphosphate soln; 10-34-0
11 Ammonium polysulfide soln; APS 20-0-40S
12 Aqua-sulfur soln
13 Ammonium thiosulfate soln; Amthio 12-0-0-26S
14 **N-pHuric® 28/27; 28-0-0-9S**
15 **N-pHuric® 15/49; 15-0-016S**
16 **N-pHuric® 10/55; 10-0-0-18S**
17 **Calcium ammonium nitrate soln; 17-0-0-8.8 Ca**
18 Iron nitrosyl; 11-0-0-7Fe
19 **Enquik®**
20 **Enzone™**
21 **Unocal Plus®**
22 **Propel®**
23 **SurpHtac II™**
24 **Nitric acid**
25 Phosphoric acid (white)
26 Phosphoric acid (green)
27 Sulfuric acid
28 Water
29 **Urea; 46-0-0**
30 **Ammonium nitrate, 34-0-0**
31 Monoammonium phosphate; MAP 10-53-0
32 Diammonium phosphate; DAP 18-46-0
33 Calcium nitrate;. 15.5-0-0-19Ca
34 Potassium chloride; 0-0-60

Bold-faced products denote Unocal Agriproducts.

Note: Compatible combinations may become incompatible under extreme temperatures and pressures. Each combination should first be evaluated on a small scale.

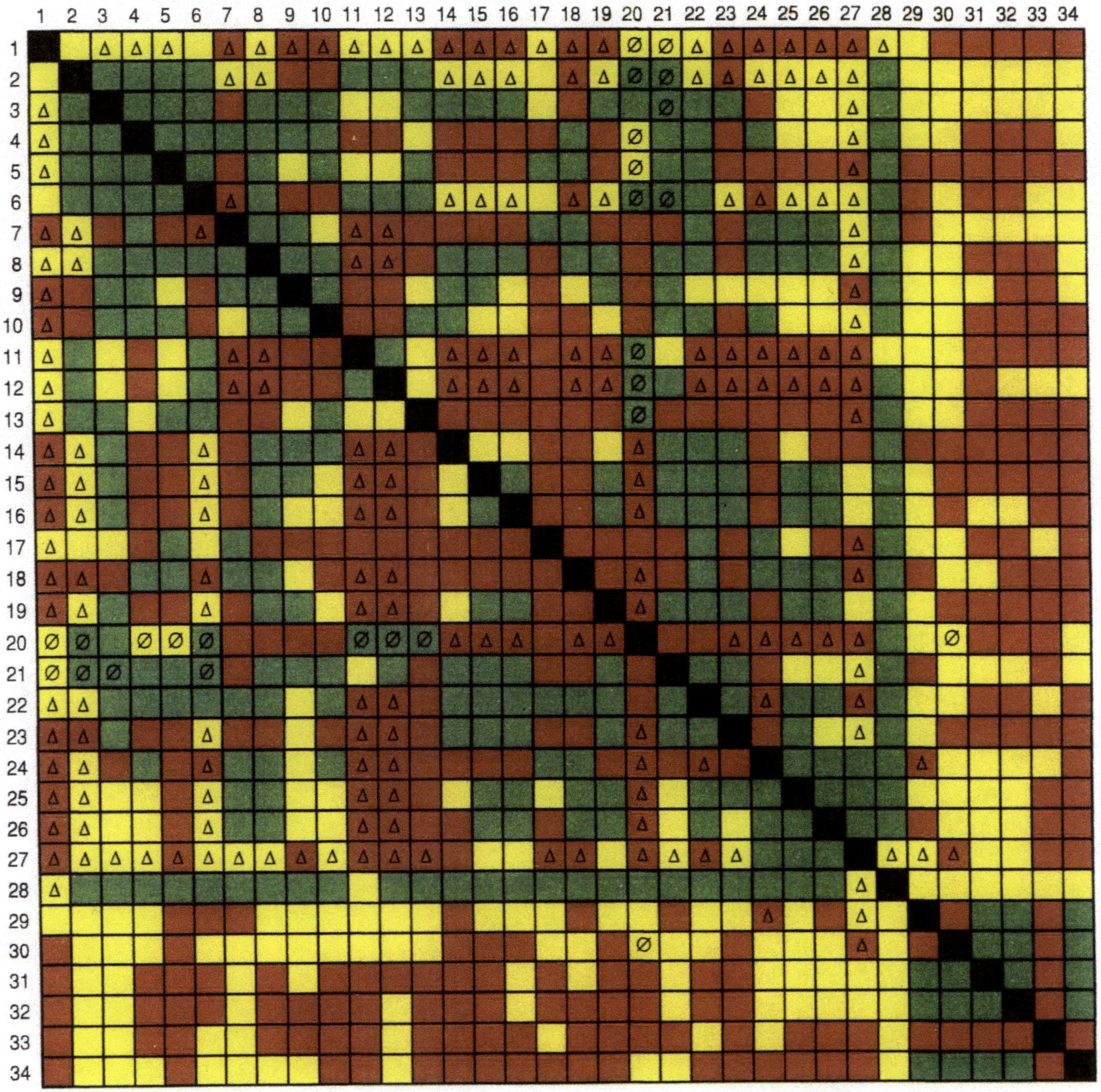

Compatible | Compatible with certain limits | Incompatible

Δ Heat generation | Ø Phytotoxic under certain conditions

Heading Off Problems

Assessing Water Quality

By Blaine Hanson, UC Irrigation and Drainage Specialist

The irrigation water to be used in a drip system should be carefully evaluated to assess any potential clogging problems. Materials such as sand, silt, and algae suspended in the water can block emitter flow passages or settle out in the drip lines wherever water velocity is low. Constituents such as calcium, bicarbonate, iron, manganese, and sulfide can also precipitate to clog emitter flow passages, and where iron and manganese concentrations are high enough, iron slimes and bacteria can grow, clogging drip lines.

Guidelines

Table 1 provides criteria developed from numerous evaluations of the effect of water quality on emitter clogging. This information can be used to assess irrigation water for clogging potential.

Table 1. Relative clogging potential of irrigation water in drip irrigation systems.

Water characteristics	*minor*	*moderate*	*severe*
Maximum suspended solids (ppm)	*<50**	*50-100*	*>100*
pH	*<7.0*	*7.0 - 8.0*	*>8.0*
Maximum total dissolved solids (ppm)	*<500*	*500-2000*	*>2000*
Maximum manganese concentration (ppm)	*<0.1*	*0.1 - 1.5*	*>1.5*
Maximum iron concentration (ppm)	*<0.2*	*0.2 - 1.5*	*>1.5*
Maximum hydrogen sulfide concentration (ppm)	*<0.2*	*0.2 - 2.0*	*>2.0*
Bacterial population (maximum number per ml)	*<10,000*	*10,000 - 50,000*	*>50,000*

** < = less than. > = more than*

Sources: D.A. Bucks, F. S. Nakayama, and R. G. Gilbert. 1979. "Trickle irrigation water quality and preventive maintenance." *Agricultural Water Management,* Vol. 2:149-62; F.S. Nakayama and D.A. Bucks. 1991. "Water quality in drip/trickle irrigation: A Review." *Irrigation Science,* Vol. 12:187-92.

Notes

1. Bicarbonate concentrations exceeding about 2 meq/liter and pH exceeding about 7.5 can cause calcium carbonate precipitation.
2. Calcium concentrations exceeding 2.5 meq/liter can cause precipitates to form during injection of some phosphate fertilizers.
3. High concentrations of sulfide ions can cause iron and manganese precipitation. Iron and manganese sulfides are very insoluble, even in acid solutions.

Chemical Constituents

Irrigation water should be analyzed for the following:

1. electrical conductivity (EC)—a measure of the total dissolved salts (TDS). An approximate equation relating TDS to EC is:

$$\text{TDS (ppm)} = 640 \times \text{EC (dS/m or mmhos/cm)}$$

2. pH
3. calcium (Ca)
4. magnesium (Mg)
5. sodium (Na)
6. chloride (Cl)
7. sulfate (SO_4)
8. carbonate/bicarbonate (CO_3 / HCO_3)
9. iron (Fe)
10. manganese (Mn)

Units of Measurement

The most common measurement unit for reporting concentrations is parts per million (ppm). Concentrations are also reported as milligrams per liter (mg/*l*). For practical purposes, ppm equals mg/*l* for irrigation water.

Concentrations may be reported in kilograms per cubic meters (kg/m^3) which is the SI unit. Kg/m^3 is the same as mg/*l*.

Concentrations may also be reported in milliequivalents per liter (meq/*l*). Conversion factors are needed to convert from mg/*l* to meq/*l* and vice versa. Table 2 list conversion factors for common constituents.

Grains per gallon may be used as a concentration unit. To convert grains per gallon to mg/*l*, multiply the grains per gallon by 17.12.

Table 2. Conversion factors: parts per million and milliequivalents per liter.

constituent	convert ppm to meq/*l*	convert meq/*l* to ppm
	multiply by	
Na (sodium)	0.043	23
Ca (calcium)	0.050	20
Mg (magnesium)	0.083	12
Cl (chloride)	0.029	35
SO_4 (sulfate)	0.021	48
$C0_3$ (carbonate)	0.033	30
HCO_3 (bicarbonate)	0.016	61

Examples:

1. convert 415 ppm of Na to meq/*l*:

$$meq/l = 0.043 \times 415 \text{ ppm} = 17.8$$

2. convert 10 meq/*l* of SO_4 to ppm:

$$ppm = 48 \times 10 \text{ meq}/l = 480$$

The quality of the data should be evaluated using the following procedures:

a. The sum of the cations (Ca, Mg, Na), expressed in milliequivalents per liter (meq/l) should about equal the sum of the anions (Cl, CO_3, HCO_3, SO_4). If the sums are exactly equal, then one of the constituents was found by differences.

b. The sum of the cations and the sum of the anions should each equal about 10 times the EC.

If these procedures reveal poor quality, the chemical analysis should be repeated.

Evaluating Your Water Quality

The following steps are guidelines for evaluating your water quality. Refer to Table 1 to assist you.

1. What is the total dissolved solids concentration? If the electrical conductivity is given only, multiply this EC (mmhos/cm) by 640 to determine the total dissolved solids.
2. What is the calcium concentration? If the calcium concentration exceeds 2.5 meq/*l*, read the chapter "Fertilizer-Irrigation Water Interactions".
3. What is the bicarbonate concentration? If the bicarbonate concentration exceeds about 2 meq/*l*, read the chapter "Clogging: An Overview" and "Calcium Carbonate Clogging".
4. What is the iron and manganese concentrations? If either concentration exceeds about 0.2 ppm, read the chapter "Clogging: An Overview" and "Iron and Manganese Clogging".

Hardness and Alkalinity

The hardness and alkalinity of a water may be reported for a water analysis, although these characteristics normally are not used for assessing potential clogging problems in drip irrigation.

The hardness of the water is primarily due to calcium and magnesium ions. Hard water will tend to precipitate calcium carbonate. Thus, the higher the hardness, expressed in terms of calcium carbonate, the higher the potential for calcium carbonate precipitation in drip irrigation systems. Classifications of hardness are:

0-75 mg/*l* - soft
75-150 mg/*l* - moderately hard
150-300 mg/*l* - hard
more than 300 mg/*l* - very hard

Alkalinity of a water is a measure of its ability to neutralize acids. Alkalinite is caused mostly by carbonate, bicarbonate and hydrology ions. Lowering the pH of waters with a large alkalinity will require more acid than for waters with a lower alkalinity.

Examples

Table 2 gives water quality data from the analysis of two irrigation water samples. *Examples 1* and *2* use the water quality data from Table 2 to evaluate the clogging potential of these irrigation waters.

Table 2. Water quality analysis of two irrigation water samples.

Water 1	*Water 2*
EC = 2.51 dS/m (1900 ppm)[1]	*EC = 0.87 dS/m (560 ppm)*[2]
pH = 7.4	*pH = 7.7*
Ca = 13.3 meq/l	*Ca = 1.9 meq/l*
Mg = 10.1 meq/l	*Mg = 1.3 meq/l*
Na = 5.4 meq/l	*Na = 5.5 meq/l*
Cl = 4.5 meq/l	*Cl = 2.0 meq/l*
HCO_3 *= 5.2 meq/l*	HCO_3 *= 2.0 meq/l*
SO_4 *= 19 meq/l*	SO_4 *= 4.7 meq/l*
Mn= less than 0.1 ppm	*Mn= 2.6 ppm*
Fe = less than 0.1 ppm	*Fe = 0.65 ppm*

[1] *Total dissolved salts = 757 x EC*
[2] *Total dissolved salts = 644 x EC*

Example 1. The relatively high total dissolved salts (1900 ppm) indicates that *Water 1* has some clogging potential. This is verified by the relatively high bicarbonate concentration (5.2 meq/l), compared to the standard of 2.0 meq/l. The calcium concentration and the bicarbonate concentration together suggest that calcium carbonate could clog the emitters, particularly if the pH were to rise as a result of any chemical injection. The iron and manganese concentrations indicate little potential for clogging from precipitation of those elements.

Example 2. The analysis of *Water 2* reveals little potential for clogging from total dissolved salts (560 ppm), but the pH and bicarbonate concentrations indicate that clogging might result from calcium carbonate precipitation. The levels of manganese and iron indicate a severe potential for clogging from manganese oxide precipitation and iron oxide precipitation.

References

Nordell, E. 1961. *Water Treatment for Industrial and Other Uses.* Reinhold Publishing Corporation, N.Y. 598 p.

Barnes, D. and F. Wilson. 1983. *Chemistry and Unit Operations in Water Treatment.* Applied Science Publishers, London and N.Y. 325 p.

Sawyer, C.N. and P.L. McCarty. 1967. *Chemistry for Sanitary Engineers.* McGraw-Hill Book Company, N.Y. 518 p.

Maintaining the Drip System

By Larry Schwankl, UC Irrigation Specialist
and Terry Prichard, UC Water Management Specialist

To work effectively, a drip system must be well-maintained. Although an automated drip system requires minimal labor for irrigation purposes, substantial labor may be needed to keep the system working at optimum efficiency and uniformity.

Routine Maintenance

Routine maintenance can include checking for leaks, backwashing filters, periodically flushing lines, chlorinating, acidifying, and cleaning or replacing clogged emitters.

Filters

Filters — whether screen, disk, or media — should be backwashed periodically to clear any collected particulate or organic matter, since clogged filters can reduce pressure to the system, lowering the water application rate. Screens should be cleaned when the pressure loss between screen inlet and outlet increases by 3 to 5 psi. Sand media filters should be backwashed when the pressure drop across the filters is about 5 to 10 psi. With either type of filter, backwashing can be done either manually — by removing the screen and physically cleaning it — or automatically. Screen filters can be cleaned automatically either on a regular schedule or when the pressure differential across the filter becomes unacceptably high — greater than 4-6 psi. A clogged filter will not pass the necessary irrigation water and the pressure differential across the clogged filter may lead to filter failure.

For backwashing purposes, screen types are *back-flushing* (where the flow of water is reversed for cleaning), *blowdown* (where a high velocity jet of water is run over the screen to wash away contaminants), or *gravity-flow,* (where water is discharged onto the screen by gravity before water is pumped into the irrigation system). Media filters can be backwashed by reversing the flow of water through the filter media so that the media particles are suspended and agitated. This upward movement of water promotes the release and flushing of contaminants from the system. While one of the media filters is backwashing, the other filter pod(s) increases its filtration rate to compensate. Recommended backwashing flowrates are 10 to 15 gallons per minute per square foot of filter area for sand sizes of 20 and 30, and 20 to 25 gallons per square foot for sand sizes of 11 and 16.

Disc filters are backflushed by loosening the discs and allowing water to flow in the opposite direction. One manufacturer uses a specially designed nozzle system to spin the discs during backflushing. Manufacturers claim that disc filters require less water for backflushing than sand media filters.

Flushing Lines

Periodically flushing the lateral lines—opening the lines and allowing them to flush clear—is essential, since the filters trap only the large contaminants entering the system. The lateral lines therefore collect material that may eventually clog the emitters. Flushing clears the system of many contaminants.

How often the laterals should be flushed depends on the quality of the irrigation water and on the degree of filtration. Generally, flushing should be performed bi-weekly, although less frequent flushing may be found to be adequate. The laterals should also be flushed following fertilizer or chemical injection and any periodic chlorine injection.

Automatic flushing valves are also available. These valves, installed on the ends of the laterals, remain open until enough pressure builds up in the system to close them. Automatic flushing valves work well in some systems, but under certain conditions do not provide a long enough flushing period to rid the lateral lines of all contaminants.

Chlorination

For waters with a high organic load (algae, moss, bacterial slimes), chlorination should be done with chlorine gas, sodium hypochlorite, or calcium hypochlorite. Whether chlorination should take place continually (1 to 2 ppm free chlorine at the end of lateral line) or periodically (approximately 10 ppm free chlorine at lateral end) depends on the severity of the clogging. Continual chlorination is usually necessary where the clogging potential is severe.

Surface water sources are more likely than groundwater sources to contain organic substances that can cause clogging. Well water pumped into and stored in a pond or reservoir should be considered a surface water source.

Acidification

Acidification—adding acid—may be necessary for irrigation water that has a tendency to form chemical precipitates (lime or iron). Groundwater sources are more susceptible than surface water to chemical precipitation. The chapter, "Clogging: An Overview," page 139 in this manual, discusses chemical precipitation.

Acidifying to lower the pH of the water to 7.0 or below will usually be sufficient to minimize chemical precipitate problems. Acids that can be added to the irrigation water include sulfuric, hydrochloric, or phosphoric acid. A nitrogen fertilizer/sulfuric acid mix is frequently used and is safer to handle than acid. Acidification has the added benefit of increasing the efficacy of chlorine additions.

Emitter Maintenance

Emitters may have to be cleaned or replaced because of clogging. The laterals and emitters should be inspected routinely to identify drip emitters that are completely clogged, although this will probably not identify emitters in which the flow has been only reduced. Partially clogged emitters will be located only if water is collected from the emitters to determine their discharge rate.

The first step in cleaning emitters is to determine what is causing the clogging. Emitters and material caught when the laterals are flushed can be examined to determine the cause. If organic matter is the culprit, a high level of chlorine (50-100 ppm) can be injected into the line and allowed to sit for about twenty-four hours. The lines should then be thoroughly flushed. If chemical precipitation is causing the clogging, acid can be injected to lower the pH to approximately 5.0. The acid should be allowed to sit in the line for twenty-four hours and then flushed. If this does not clear the emitters, the drip tape will have to be replaced.

If pressure-compensating emitters are being used, the manufacturer should be contacted to determine the minimum pH allowable. The flexible orifice in some pressure-compensating emitters will be damaged by water with a pH of 4.0 or below.

Periodic Maintenance

Maintenance tasks that should be carried out on a less-frequent basis include inspecting the filter media, inspecting the pressure-regulating valve, and replacing pressure gauges.

Filter media tend to cake together over time and as a result may fail to provide good filtration. Frequent backwashing may be symptomatic of such a problem. Sand media should be replaced if this occurs. When the old media is removed, the underdrain system should be inspected. Even if the sand media appears to be in good condition, additional media may be added periodically, since some of the sand is invariably lost during the backwash cycle.

Adjustable pressure-regulating valves, set at installation, should be inspected and adjusted periodically to see that correct operating pressure is maintained. Pre-set pressure-regulators should be inspected to ensure that they are operating properly. Foreign material in the line may jam the adjustment mechanism and inhibit operation.

Pressure gauges tend to eventually wear out and should be replaced if the accuracy is in question. Liquid-filled pressure gauges, which are slightly more expensive, may be a good replacement choice. Gauges must be scaled to operate in a pressure range appropriate to the system.

Leaks

Drip systems should be inspected regularly for leaks, which can occur in compression fittings, end closures, or hose adapters. Rodent and insect damage to drip tapes can cause severe leaks.

Clogging: An Overview

By Larry Schwankl, UC Irrigation Specialist
Blaine Hanson, UC Irrigation and Drainage Specialist
and Terry Prichard, UC Water Management Specialist

Precipitating chemicals and organic contaminants can clog drip irrigation systems. When a drip system using groundwater for irrigation becomes clogged, the usual cause is chemical precipitation from calcium carbonate (lime), iron, or manganese in the irrigation water.

Lime Precipitation

Calcium carbonate (lime) precipitation is the most common cause of chemical clogging in drip irrigation. Water with a pH of 7.5 or above and bicarbonate levels of 2 meq/l (100 ppm) is susceptible to lime precipitation if comparable levels of calcium are present naturally in the system or if a compound containing calcium is injected into the system.

The usual treatment for lime precipitation is to acidify the water to lower the pH to 7.0 or below. Litmus paper, colormetric kits, or portable pH meters can be used to determine the water's pH. Sulfuric acid is usually used to reduce pH, but phosphoric acid and hydrochloric acid may also be used. Since handling acids is hazardous, some water managers prefer to use one of the safer acid/fertilizer compounds now available. Researchers are evaluating other compounds — including a phosonate material and several polymer materials — to determine their efficacy in preventing calcium carbonate precipitation.

Iron and Manganese Precipitation

Iron and manganese precipitation can cause clogging even at low concentrations: iron: 0.2 ppm or greater, manganese: 0.15 ppm or greater. These compounds, which are most often present in groundwater, are in a soluble reduced state in the well, but oxidize and precipitate as very small but solid particles when exposed to the atmosphere. Iron and manganese will precipitate across a wide range of pH (iron, for example, will precipitate at pH 6-9.5), which include the pH levels of almost all naturally occurring waters.

Iron precipitate is characterized by a reddish stain and rust particles in the water. Manganese precipitate has a similar appearance, but the stain is darker — nearly black in color.

Iron/manganese precipitation is further complicated by bacteria that use iron/manganese as energy sources. These bacteria form filamentous slimes that can clog filters and emitters and that can also provide the matrix or glue to hold other contaminants in the system. Iron bacteria can be controlled by injecting chlorine continually at 1-2 ppm residual (at the end of the line) or intermittently at 10-20 ppm residual for about one hour.

How to Mitigate Chemical Iron or Manganese Precipitation

The following measures can be taken to mitigate chemical iron or manganese precipitation:

1. Aeration and settling. Water can be pumped into a pond or reservoir and allowed to aerate from contact with the atmosphere. The iron precipitate is then allowed to settle out. Additional aeration of the water may be necessary to ensure that the iron is oxidized. After the iron settles, the water can be drawn off for use.

2. Chlorine precipitation and filtration. Injecting chlorine into the water will oxidize the dissolved (ferrous) iron, causing it to precipitate. The precipitated iron (ferric oxide) can then be filtered out, preferably with a sand media filter, which can be readily backwashed.

3. pH control. Where the potential for iron precipitation exists, lowering the pH in the system to less than 4.0 will keep the iron from precipitating. The cost of this practice may limit its use.

4. Chelation. In municipal water treatment, a polyphosphate, such as sodium hexametaphosphate, is added to the water before the iron is oxidized. This prevents agglomeration of the small individual particles. Recommended injection rates are 2 mg/l of sodium hexametaphosphate for each 1 mg/l of iron or manganese. Since this practice is expensive it should only be used in agricultural systems after careful evaluation.

Miscellaneous Compounds

Other compounds that can cause clogging include magnesium carbonate, calcium sulfate, and zinc injected in sulfate form. Adding anhydrous or aqua ammonia to irrigation water will increase its pH, possibly facilitating the precipitation of calcium or magnesium compounds. Adding phosphate fertilizers may also cause the phosphate to react with calcium or magnesium, resulting in a precipitate. This can be prevented by adding acid to significantly lower the pH of the water.

Table 1 summarizes recommended treatments for various types of chemical and biological clogging.

Table 1. Water treatments to prevent clogging in drip irrigation systems.

Problem	Treatment Options
Carbonate precipitation (white precipitate) HCO_3 greater than 2.0 meq/l pH greater than 7.5	1. Continuous injection: maintain pH between 5 and 7 2. Slug injection: maintain pH at under 4 for 30-60 minutes daily
Iron precipitation (reddish precipitate) Iron concentrations greater than 0.1ppm	1. Aeration and settling to oxidize iron. (Best treatment for high concentrations—10 ppm or more.) 2. Chlorine precipitation—injecting chlorine to precipitate iron: a. use an injection rate of 0.7 ppm of chlorine per 1 ppm of iron b. maintain a pH of 7. c. inject in front of the filter so that the precipitate is filtered out 3. Reduce pH to 4 or less to prevent precipitation. 4. Inject inhibitors to prevent precipitation.
Manganese precipitation (black precipitate) Manganese concentrations greater than 0.1 ppm	1. Inject 1.3 ppm of chlorine per 1 ppm of manganese in front of front of the filter. Maintain a pH of at least 9.5.
Iron bacteria (reddish slime) Iron concentrations greater than 0.1 ppm	1. Inject chlorine at a rate of 1 ppm free chlorine continuously or 10 to 20 ppm for 30 to 60 minutes daily.
Sulfur bacteria (white cotton-like slime) sulfide concentrations greater than 0.1 ppm	1. Inject chlorine continuously at a rate of 1 ppm per 4 to 8 ppm of hydrogen sulfide, or 2. Inject chlorine intermittently at 1 ppm free available chlorine for 30 to 60 minutes daily.
Algae, slime	1. Inject chlorine at a rate of 1 to 2 ppm continuously or 10-20 ppm for 30 to 60 minutes at the end of each irrigation cycle.
Iron sulfide (black sand-like material) Iron and sulfide concentrations greater than 0.1 ppm	1. Dissolve iron by injecting acid continuously to lower pH to between 5 and 7.

Calcium Carbonate Clogging

By Blaine Hanson, UC Irrigation and Drainage Specialist

Precipitation of calcium carbonate is a common problem causing clogging in drip irrigation systems. Magnesium carbonate clogging is much less of a problem because its solubility in water is at least 20 times greater than that of calcium carbonate. Reasons for carbonate precipitation include the following:

1. Change in the pH of groundwater. This can occur when groundwater is pumped. Pumping reduces the pressure of the groundwater as it flows into the well. This reduction releases dissolved carbon dioxide gas causing the pH of the groundwater to increase. The pH increases can result in carbonate precipitation.
2. Evaporation of water in the drip emitter or microsprinkler. Evaporation increases the concentration of chemicals dissolved in the water remaining in the emitter. Because of its low solubility in water, calcium carbonate readily precipitates during evaporation.
3. Increase in water temperature. The solubility of calcium carbonate is reduced as the water temperature increases.
4. Injection of certain chemicals such as aqua ammonia and bleach chlorine.

The occurrence of calcium carbonate in water depends on the pH and the temperature. Figure 1 shows the pH dependence of the various species of carbon dioxide. At pH values less than about 6, mostly dissolved carbon dioxide and a small amount of carbonic acid exist in the water. At pH values between about 6.5 and 10, bicarbonate is the dominate species. At pH values greater than about 10.5, the carbonate ion dominates.

This behavior indicates that lowering the pH of the water will prevent or reduce carbonate clogging of drip irrigation systems. Lowering the pH dissolves any existing carbonate precipitation and prevents the formation of carbonate.

Potential Problem and Recommendations

Acid injection is used to lower the pH of the water. Common acids used include sulfuric acid and murietic acid or hydrochloric acid. Other acids such as citric acid and nitric acid can be used but they are more expensive. An acid/fertilizer compound such as N-pHRIC can also be used, which is safer to use than straight acids.

The potential for carbonate clogging exists when the bicarbonate concentration in the water exceeds 2 meq per liter and the pH exceeds 7.5, based both on theoretical approaches and on the experience of the drip irrigation industry. Recommended treatments for preventing clogging are:

1. Inject acid continuously to maintain the pH between 5 and 7.

2. Inject acid intermittently to maintain the pH at under 4.0 for 30-60 minutes. The injection frequency will depend on the rate at which carbonate precipitation occurs.

How Much Acid is Needed

The amount of acid needed to lower the pH to a desired level depends on the bicarbonate/carbonate concentration in the water and the target pH. Two approaches can be used to determine the amount of acid needed:

1. Develop a titration curve which shows the amount of acid needed to lower the pH to a given value. This is usually done by a laboratory, and it requires a sample of both the water and the acid. Figure 2 shows the titration curve using N-pHURIC 49/15 to lower the pH.

2. Use a trial and error approach by adding acid to the water in increments and measuring the pH until the desired pH is reached.

Injection Rate

When the amount of needed acid is determined, the injection rate of the acid can be calculated by the following equation.

$$\text{Injection Rate (gallons per hour)} = \frac{A \times Q \times 60}{326{,}000}$$

where A is gallons of acid per acre foot of water needed to lower the pH and Q is the system flowrate in gallons per minute.

Example

Calculate the injection rate needed to lower the pH to 4 using N-pHURIC (15/49). Figure 2 is the titration curve. The system flowrate is 1000 gallons per minute.

a. From Figure 2, 26 gallons of N-pHURIC per acre-foot of water are needed to lower the pH to 4.

b. Injection Rate = 26 gallons / acre-foot x 1000 gpm x 60 / 326,000
= 4.8 gallons per hour.

Measuring the pH

The pH can be measured in the field by using litmus strips of paper, pocket pH meters, or test kits for aquariums or swimming pools. These test kits are readily available in stores selling swimming pool supplies and so forth. Pocket pH meters can be purchased from companies such Ben Meadows Company (800-241-6401). However, these meters require periodic calibration using standard solutions that can also be purchased.

Caution

1. Do not mix acid and chlorine. This causes chlorine gas to form.
2. Do not add acid to water.

Amendments

Several amendments are being recommended for preventing carbonate precipitation. However, little information on the effectiveness of these amendments exists at this time.

References

Barnes, D. and F. Wilson. 1983. Chemistry and Unit Operations in Water Treatment. Applied Science Publishers, London and NY. 325 p.

Nordell, Eskel. 1961. Water Treatment for Industrial and Other Uses. Reinhold Publishing Corp. NY. 598 p.

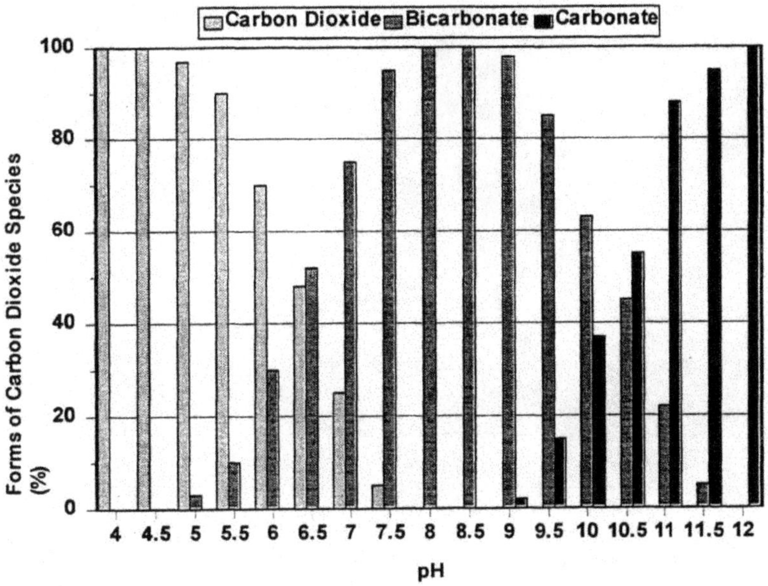

Figure 1. pH versus amounts of carbon dioxide, bicarbonate, and carbonate.

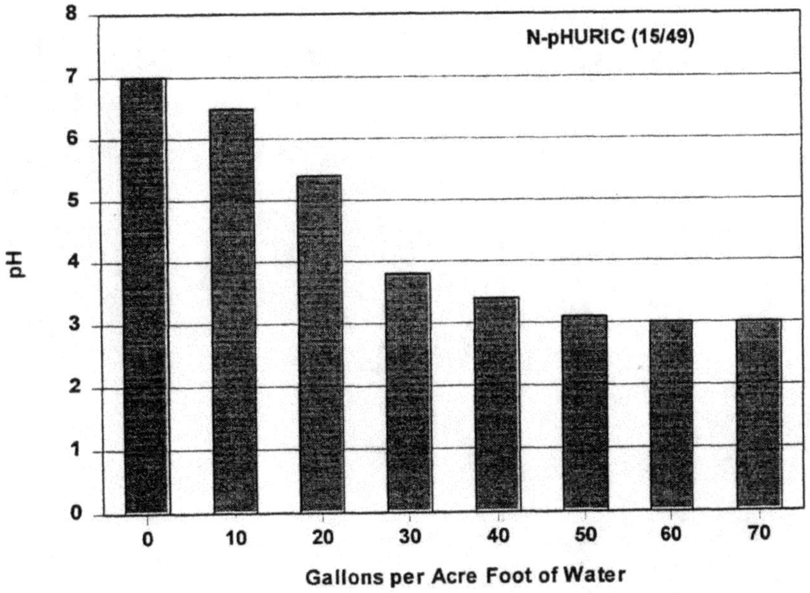

Figure 2. Titration curve using N-pHURIC (15/49)

Iron and Manganese Clogging

By Blaine Hanson, UC Irrigation and Drainage Specialist

Iron Chemistry

Iron frequently is found in groundwater, the result of groundwater contact with rocks and other aquifer material containing iron. Water contact with metal well casings, pump parts, and steel pipelines can also dissolve iron.

In groundwater where dissolved oxygen concentrations are low and the pH is less than about 6, iron exists as ferrous iron. This form of iron, normally ferrous bicarbonate, is soluble in water and does not cause emitter clogging in drip irrigation systems.

Pumping groundwater lowers the water pressure which releases dissolved carbon dioxide gas thus raising the pH, raises the water temperature and exposes the groundwater to air. Aeration of the groundwater can occur from cascading water in the well or leaks on the suction side of a centrifugal pump used for pumping shallow groundwater. These changes can cause the ferrous iron to become ferric iron, a process called oxidation. At pH's less than about 5, the oxidation rate is very slow compared to that at higher pH's, and thus little ferric iron is formed. Ferric iron is insoluble in water that is alkaline or weakly acidic. Aerated water with a pH between 7 and 8.5 contains mostly insoluble ferric iron. This iron precipitates out of the water as iron oxides, a reddish-brown deposit. This form of iron causes clogging problems in drip irrigation systems. Concentrations as low as 0.2 ppm can cause clogging problems.

Iron Bacteria

Frequently, bacteria are found when iron is present in the water. Sources of the bacteria are unclear, but bacteria may be the result of well contamination during drilling. Most of these bacteria derive their energy from the oxidation of soluble ferrous iron to insoluble ferric hydroxide. This process results in iron precipitation along with the formation of a slime inside pipelines, emitters, and so forth.

Manganese Chemistry

Manganese resembles iron in its chemical behavior, but it is less abundant than iron. It can occur as a soluble manganous manganese or manganese bicarbonate which changes to an insoluble manganese hydroxide when it reacts with atmospheric oxygen. Manganese hydroxide precipitates out as a black deposit. Concentration as low as 0.1 ppm can cause clogging problems. However, the oxidation rate of manganese is very slow at pH's less than about 9.5. Most irrigation waters have a pH much less than 9.5, thus, clogging caused by manganese precipitation is less of a problem compared with iron clogging.

Clogging Prevention

Methods for preventing clogging by iron and manganese precipitation are as follows:

1. Aerated the water to oxidize the iron or manganese and allow the precipitate to settle out before using the water for irrigation. This method requires a reservoir or settling basin with a maximum water velocity of about 1 foot per second. Chlorine may be needed to control any algae growth in the reservoir or basin. This may be the only practical method for dealing with high iron concentrations. The effect of the aeration is:

a. Causes oxygen to be dissolved into the water.
b. Releases dissolved carbon dioxide gas, thus increasing the pH.
c. Releases or partially releases other dissolved gases such as hydrogen sulfide.

The oxidation rate under aeration depends on the pH of the water and the contact time. At a pH of 5, the oxidation rate is slow with little of the ferrous iron being oxidized according to one experiment (Table 1). Irrigation water with a pH of 5 to 5.5 and iron concentrations of 2 to 4 ppm has been used with little or no clogging problems. At a pH of 7, almost all of the ferrous iron was oxidized by air after 15 minutes of contact time. About 1 ppm of dissolved oxygen will oxidize about 7 ppm of ferrous iron.

Table 1. Effect of pH and contact time on the oxidation of ferrous iron.

Raw Water		Iron in Aerated and Filtered Water (ppm)		
pH	Iron (ppm)	(15 min.)	(30 min.)	(60 min.)
5.0	10	9.0	-	7.5
5.5	10	5.5	4.6	4.0
5.95	10	5.0	4.0	3.5
6.15	10	4.4	3.5	2.5
6.5	10	2.8	1.8	0.3
6.65	10	0.7	0.2	0.1
6.8	10	0.2	0.1	<0.1
7.0	10	0.1	<0.1	<0.1
7.45	10	0.1	<0.1	<0.1
8.05	10	<0.1	<0.1	<0.1

(Source: Nordell, Eskel. 1961. *Water Treatment for Industrial and Other Uses*. Reinhold Publishing Corp., NY. 598 p.)

Theoretically, about 1 ppm of dissolved oxygen will also oxidize about 7 ppm of manganese. However, the pH must be greater than 9.5 for any substantial oxidation to occur as shown in Table 2.

Table 2. Effect of pH and contact time on the oxidation of manganous manganese.

Raw Water		Manganese in Aerated and Filtered Water (ppm)		
pH	Mn (ppm)	(15min.)	(30min.)	(60min.)
8.5	10	-	-	10.0
9.0	10	-	-	9.0
9.3	10	8.5	8.0	7.5
9.5	10	7.5	5.0	3.2
9.7	10	3.0	1.3	0.9
9.95	10	0.9	0.7	0.6
10.3	10	<0.02	<0.02	<0.02

2. Inject chlorine to oxidize the iron or manganese and cause precipitation. Injection must occur upstream of the filter to filter out the precipitate. The rate at which chlorine is mixed in the water is also a factor in the process. Centrifugal sand separators installed immediately downstream from the injection point have been used to increase the mixing.

An injection rate of 1 ppm of chlorine will oxidize about 1.6 ppm of ferrous iron, or conversely, inject about 1.0 ppm of chlorine for each 0.7 ppm of ferrous iron. However, chlorine oxidizes the iron at a lower pH than does aeration. In one experiment, almost all of the iron (initial concentration of 10 ppm) was oxidized at a pH of 5.0 for a 15 minute contact time. For practical purposes, however, a pH of about 6.5 to 7.5 is recommended for chlorine oxidation.

Inject about 1.3 ppm of chlorine to oxidize each 1.0 ppm of manganese. However, the pH must be at least 9.5 for any appreciable oxidation to occur. One experiment showed, that at a pH of 9.0, very little oxidation occurred, whereas, most of the manganese was oxidized by chlorine at a pH of 9.5.

3. Decrease the pH to 4 or less to prevent precipitation. At a low pH, the oxidation rate of ferrous iron to ferric iron is very slow. Little clogging problems has been reported for iron concentrations of 2 to 4 ppm at a pH's of 5 to 5.5. This practice, however, can be expensive and may damage the irrigation system.

4. Inject an inhibitor that prevents iron or manganese precipitation. Compounds such as polymeric acid, polyphosphate, or phosphonic acid have been

used as inhibitors. These compounds must be injected prior to any oxidation of ferrous iron to ferric iron. These materials will not work if ferric iron has already formed.

NOTE: Once the iron has precipitated, a very low pH (1 or 2) is needed to dissolve the precipitate. These low pH's can damage components of drip irrigation systems.

Controlling Iron Bacteria

Chlorine is recommended for controlling iron bacteria. Inject chlorine to continuously to maintain a free chlorine concentration of 1 ppm or maintain a concentration of 10 to 20 ppm daily for 30 to 60 minutes. Since bacteria may be growing in the well, chlorine should be injected into the well to control these bacteria.

Iron Sulfide

Iron sulfide can form where hydrogen sulfide is found in the water. Iron sulfide precipitate is a black precipitate. Methods for preventing iron sulfide precipitation are the same as for iron hydroxide precipitation. Hydrogen sulfide concentrations greater than 0.2 ppm can cause clogging problems. Iron sulfide is very insoluble even in acid solutions.

References

Barnes, D. and F. Wilson. 1983. *Chemistry and Unit Operation in Water Treatment.* Applied Science Publishers, London and NY. 325 p.

Keller, J. and R. D. Bliesner. 1990. *Sprinkle and Trickle Irrigation.* VanNostrand Reinhold, NY. 652 p.

Nordell, Eskel. 1961. *Water Treatment for Industrial and Other Uses.* Reinhold Publishing Corp., NY. 598 p.

Chlorination

By Larry Schwankl, UC Irrigation Specialist,
Blaine Hanson, UC Irrigation and Drainage Specialist
and Terry Prichard, UC Water Management Specialist

Chlorine is often added to irrigation water to oxidize and destroy biological micro-organisms such as algae, fungi, and bacteria. While these micro-organisms may be present in water from any source, they are most likely to be present at high levels in surface water from rivers, canals, reservoirs, and ponds.

When water containing high levels of micro-organisms is introduced into a drip irrigation system, emitters can become clogged. Using good filters (such as media filters) and acidifying the water can cut down on organic clogging, but the best way to deal with the problem is to add a biocide such as chlorine.

Dissolving chlorine in water produces hypochlorous acid, which becomes ionized, forming an equilibrium between the hypochlorous acid and hypochlorite — referred to collectively as the *free available chlorine.* Hypochlorous acid is a more powerful biocide than hypochlorite. Acidifying the water tends to favor the production of hypochlorous acid and thus makes the chlorine added more effective. It is important to NOT mix chlorine and acids together, since this causes the formation of chlorine gas, which is toxic.

The most common sources of chlorine are sodium hypochlorite (a liquid), calcium hypochlorite (powder or granules), and chlorine gas.

Hypochlorite

Sodium hypochlorite is usually available with up to 15% available chlorine. Household bleach is sodium hypochlorite, with 5.25% active chlorine. To determine the chlorine injection rate when using sodium hypochlorite, the following formula can be used:

Calcium hypochlorite with 65-70% available chlorine can usually be obtained. In using the formula given above, note that 12.8 pounds of calcium hypochlorite added to 100 gallons of water will form a 1% chlorine solution. A 2% chlorine solution would therefore require adding 25.6 pounds of calcium.

The injection rate needed for these materials is calculated with the following equation:

$$\text{Chlorine injection rate (gal/hr)} = \text{System flow rate (gpm)} \times \text{Desired chlorine concentration (ppm)} \times 0.006 \div \text{Strength of chlorine solution (\%)} \qquad (1)$$

hypochlorite to 100 gallons of water. Any chlorine stock solution can be mixed following the same pattern.

Chlorine Gas

Chlorine gas contains 100% available chlorine. While using chlorine gas is generally considered the least expensive method of injecting chlorine, it is also the most hazardous and requires extensive safety precautions. The chlorine gas injection rate can be determined from the following formula:

$$\text{Chlorine gas injection rate (lbs/day)} = \text{System flow rate (gpm)} \times \text{Desired chlorine concentration (ppm)} \times 0.012 \quad (2)$$

Levels of Treatment

If the irrigation water has high levels of algae and bacteria, continuous chlorination may be necessary. The recommended level of free available chlorine is 1 to 2 ppm, measured at the end of the farthest lateral with a good quality chlorine test kit.

Periodic injection (once every two to three weeks), at a higher chlorine rate (10-20 ppm) may be appropriate where algae and bacterial slimes are less of a problem. The frequency of the chlorine injection depends on the extent of organic clogging.

Superchlorination—bringing chlorine concentrations to within 500 to 1000 ppm—is recommended for reclaiming drip irrigation systems clogged by algae and bacterial slimes. Superchlorination requires special care to avoid damage to plants and to irrigation components.

Considerations

The following should be considered in performing chlorination:

- Injection of sodium hypochlorite and calcium hypochlorite will increase the pH of the water. If the pH exceeds 7.5 and at least 2 milliequivalent per liter of bicarbonate are in the irrigation water, calcium carbonate precipitation can occur. Use of these chlorine chemicals may require adding acid to the water to prevent the formation of carbonate.

- Injection of chlorine into the water results in hypochlorus acid and hypochlorite ions being formed. Hypochlorus acid, however, is by far the most effective chlorine form for controlling bacteria, algae, etc. The amount of hypochlorus acid in the water depends on the water's pH. The lower the pH, the more the hypochlorus acid as shown in Figure 1. Maintaining the pH at about 7 result in about 70 percent of the chlorine being in the form of hypochlorus acid.

Do not use sulfur dioxide from burning sulfur to lower the pH of the irrigation water while injecting chlorine. The reaction of the sulfur dioxide with the irrigation water forms sulfurous acid which in turn inactivates the chlorine source.

• Inject the chlorine upstream from the filter to help keep the filter clean and so that the filter can remove any precipitates that may be caused by the chlorine injection. Chlorine is a very effective oxidizing agent and will cause any iron and manganese present in the water to precipitate and to clog the emitters.

• Chlorine compounds should be stored separately in fiberglass or epoxy-coated plastic tanks. Acids and chlorine should never be stored together.

• Do not inject chlorine when fertilizers, herbicides, and insecticides are being injected, since the chlorine may destroy the effectiveness of these compounds.

• When mixing stock chlorine solutions, always add the chlorine source (dry or liquid) to the water, not vice versa.

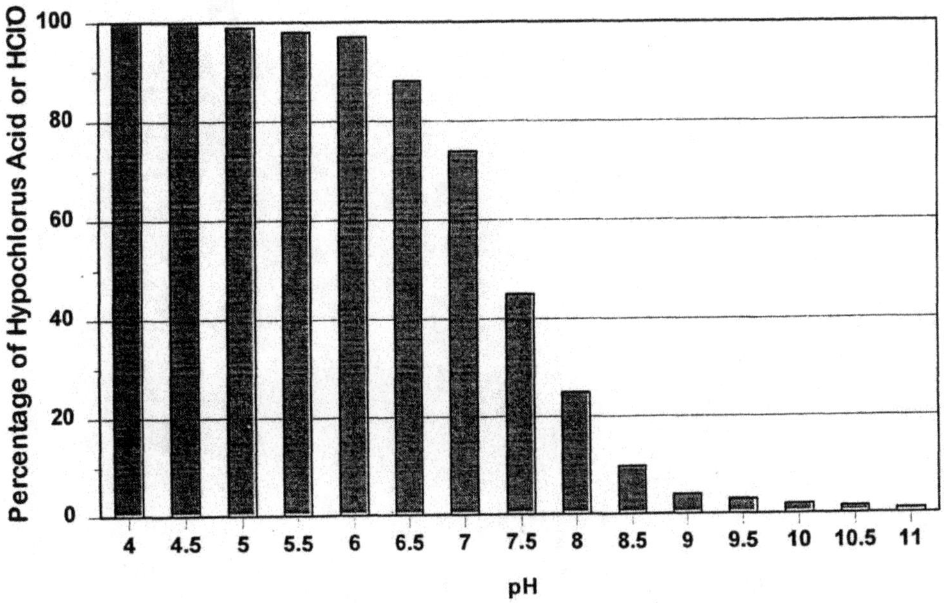

Figure 1. Percent of chlorine in the hypochlorus acid form for different levels of pH.

Calculating Injection Rates and Concentrations

By Blaine Hanson, UC Irrigation and Drainage Specialist

Acid Injection

Calculating the acid injection rate requires knowing the amount of acid needed to lower the pH to a desired level and the system flowrate. A titration curve, in Figure 1, shows the amount of acid needed to lower the pH. Developing a titration curve, done by a laboratory, requires a sample of both the irrigation water and the acid.

The acid injection rate is calculated by:

$$IR = (A \times Q \times 60) / 326{,}000$$

where IR is the acid injection rate in gallons per hour, A is the gallons of acid needed per acre-foot of water, and Q is the system flowrate in gallons per minute.

Example

Calculate the injection rate needed to lower the pH to 4 using N-pHURIC (15/49). Figure 1 is the titration curve. The system flowrate is 1000 gallons per minute.

a. From Figure 1, 26 gallons of N-pHURIC per acre-foot of water are needed to lower the pH to 4.
b. IR = 26 gallons / acre-foot x 1000 gpm x 60 / 326,000 = 4.8 gallons per hour.

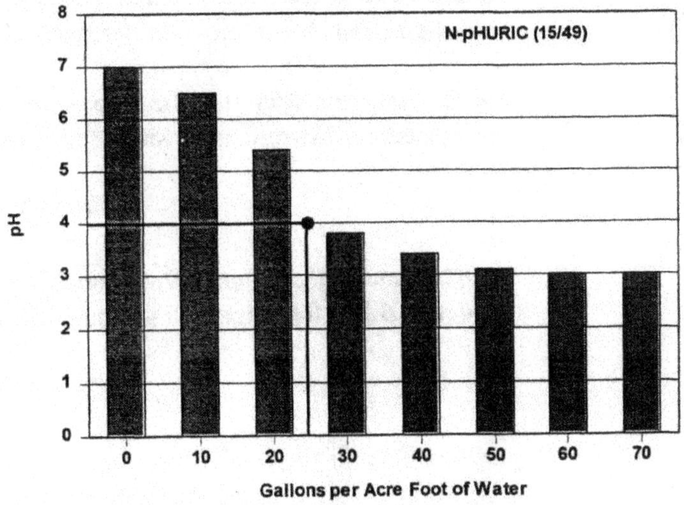

Figure 1. Titration curve.

Chlorination

Injection Rate. The chlorine injection rate can be calculated by (see chapter, "Chlorination"):

$$IR = (0.006 \times Q \times C) / S$$

where IR is the injection rate in gallons per hour, Q is the system flowrate in gallons per minute, C is the desired chlorine concentration of the irrigation water in ppm, and S is the strength of the chlorine solution (%).

Example

Calculate the injection rate needed for a chlorine concentration of 10 ppm. The system flowrate is 600 gallons per minute. A 5.25% bleach solution is used.

Concentration. Given an injection rate, strength of solution, and system flowrate, the concentration of chlorine in the irrigation water can be calculated by

$$C = (IR \times S) / (0.006 \times Q).$$

Example

Calculate the concentration of chlorine in irrigation water for an injection rate of 3 gallons per hour, a 5.25% strength, and a system flowrate of 500 gpm.

$$C = (4 \text{ gph} \times 5.25\%) / (0.006 \times 500 \text{ gpm}) = 7 \text{ ppm}$$

Strength and Dilution. Sometimes, dilution is needed to achieve the desired strength and chlorine concentration. For small system flowrates, injector rates may not be small enough to inject at a rate needed for a desired chlorine concentration, and thus dilution of the original chlorine solution may be needed. The desired strength and the amount of dilution of the original solution needed to achieve that strength can be determined by

a. Calculate the strength of the chlorine solution needed for the given injection rate, system flowrate, and desired concentration.

$$S = (0.006 \times Q \times C) / IR$$

b. Determine the gallons of water needed per gallon of the original solution needed to dilute the original solution to the desired strength.

$$V = S_1 / S_2$$

where V is the gallons of water needed to dilute one gallon of the original solution, S_1 is the strength of the original solution, and S_2 is the strength of the diluted solution as calculated in Step a.

Example

Determine the dilution needed for an injection rate of 2.5 gallons per hour, system flowrate of 200 gallons per minute, and a concentration of 2 ppm. The strength of the original solution is 5.25%.

a. S = 0.006 x 200 gpm x 2 ppm / 2.5 gph = 0.96%

b. V = 5.25% / 0.96% = 5.5 gallons. Thus 5.4 gallons of water need to be added to 1 gallons for 5.25% chlorine solution to achieve the desired strength of 0.96%.

Monitoring the System

By Blaine Hanson, UC Irrigation and Drainage Specialist

A drip system should be monitored to detect any trends in performance and to identify developing problems. If an evaluation determines that the flowrate is steadily decreasing, the cause may be declining groundwater levels, a worn pump, clogged filters or emitters, malfunctioning valves, or any combination of these factors.

Declining groundwater levels — which can be detected by periodically measuring water levels in the well — will cause both the system pressure and the flowrate to decrease. A worn pump, which will have the same effect, can be identified by conducting periodic pumping plant tests and comparing the results with the manufacturer's pump performance curve.

Clogged emitters, the most common and also the most serious problem in drip systems, can be detected by several different methods. One method is to measure the discharge rates of 10 or 20 emitters by installing containers under the emitters and measuring the volume of water in each container after a given period of time. Note that because emitter flowrates can be affected by differences in pressure as well as by clogging, the pressure must be uniform for clogging to be revealed. Pressure is usually consistent along the end of the lateral so long as the ground surface is level. This method can be used for subsurface as well as surface drip systems, but subsurface emitters must be uncovered for the test, which may disturb any clogging caused by roots and make the evaluation inaccurate.

Another way to identify clogging is to observe wetted diameters at the ground surface. Significant variation in diameters or large gaps between wetted areas may indicate emitter clogging.

Periodically measuring the system flowrate can detect clogging in both surface and subsurface drip systems. For the assessment to be accurate, the same block of laterals must be tested and the pressure downstream from the pressure-regulating valves must be the same each time the flowrate is measured, since different pressures result in different flowrates. Tests conducted on a subsurface drip system in Colorado using this procedure found a 20% decrease in the system flowrate as a result of clogging.

Reference

Mitzyed, N., and E. G. Kruse. 1989. Emitter discharge evaluation of subsurface trickle irrigation systems. *Transactions of the American Society of Agricultural Engineers,* Vol. 32: 1223-28,

Root Intrusion in Subsurface Drip Systems

By Blaine Hanson, UC Irrigation and Drainage Specialist

Plant roots tend to grow toward the soil areas with the highest water content, which in subsurface drip systems is near the emitters. Because of this tendency, roots can clog subsurface drip irrigation systems by growing into the emitter openings, especially in systems left in place for several irrigation seasons. One study conducted in Hawaii found roots plugging 1-5% of the emitters after a 22-month period.

Various strategies have been used to combat root intrusion, including water management, chemical treatment, and drip tape emitter design, but very little research has been conducted to evaluate the relative effectiveness of the approaches available. The following advice, therefore, is based more on anecdotal evidence than on direct measurement.

Water Management

Controlling the soil water content close to the emitter can discourage roots from growing in that direction. In many crops, frequent irrigations, resulting in a saturated soil near the emitter, will restrict root growth in that area and lower the potential for root intrusion. This technique is not effective for crops such as celery that thrive on very wet soil. For these crops, some growers have resorted to surface drip irrigation, while others use chemical treatment to reduce clogging from root intrusion in subsurface drip systems left in place for long periods of time.

Deficit irrigation can worsen root intrusion by attracting roots to the soil with the higher water content. Near the end of the growing season, particularly where an early irrigation cutoff date is required, preventing deficit irrigation — and combatting root intrusion — can be difficult.

Chemical Treatment

Some growers use weekly acid injections to discourage root intrusion by lowering the pH of the irrigation water to less than 4. Unfortunately, this practice can damage the emitters. Others inject chlorine at very high concentrations (100 to 400 ppm) after harvesting. Still others use N-pHURIC, continuous injection of phosphoric acid (15 ppm) or Vapam. One study on strawberries found that 5 gallons of Vapam per acre prevented root intrusion, yet did not harm the crop.

Drip Tape Design

Manufacturers promote some tape products as being less sensitive to root intrusion than others. The emitters of one product are impregnated with a chemical compound — Trifluralin — to prevent root intrusion. Others discourage root intrusion through physical design. Locating the emitter outlet away from the drip tape seam, for instance, may cut down on root intrusion because roots tend to grow along the seam. The size of the emitter outlets may also be a factor. Research on Hawaiian sugar cane found that drip tape and tubing with large emitter outlets was more susceptible to plugging from root intrusion than materials with smaller outlets.

References

Bui,. W. 1990. Performance of 'turbo model' drip irrigation tubes. *Proceedings of the Third International Irrigation Symposium,* Phoenix, AZ, Oct 28-Nov 1, 1990.

Bui, W. 1993. "Designing drip irrigation tubes for row crops. American Society of Agricultural Engineers, Paper 932131, presented at the 1993 International Summer Meeting, Spokane, Washington, June 20-23, 1993.

Preventing Backflow

By Blaine Hanson, UC Irrigation and Drainage Specialist

Groundwater can become contaminated if irrigation water containing fertilizer or other chemicals backflows into the well after the pump is stopped or if chemical injection continues after the pumps stops. Vacuum-relief valves, low pressure drains, interlocking circuits between the irrigation pump and the injection pump, and check valves installed in the pump discharge pipe and injection tubing are devices available to prevent backflow. Figure 1 shows the various components and their locations of backflow devices.

Vacuum-Relief Valve

Water flowing back down the column pipe in the well after the pump is stopped can cause a vacuum to develop in the pipeline, which can induce leaks around the check valves. A vacuum-relief valve allows air to flow into the pump discharge pipe as the water flows back down the column pipe, thereby preventing a vacuum from developing in the pipeline.

Low Pressure Drains

A low pressure drain installed upstream from the check valve intercepts any leakage from the check valve. The capacity of the drain should be adequate to intercept all of the leakage — which in one University of Nebraska study was found to be as high as 11 gallons per minute. The distance between the drain and the check valves should be sufficient to prevent leaking water from spraying over the drain.

Interlocking Circuits

An interlock between the irrigation pump and the chemical injection pump turns off the injection pump if the irrigation pump unexpectedly shuts down while chemicals are being injected.

Check Valves

A check valve should be installed upstream from the chemical injection point to prevent water stored in the mainlines and other pipelines from flowing back into the well. Studies have shown that check valves require a minimum back pressure to completely seal, and that without that minimum pressure, leaks can develop around the check valve. An inspection port should be installed just upstream from the check valve so that the seating of the valve can be monitored.

Another check valve should be installed in the injection pump discharge pipe to prevent irrigation water from flowing into the injection system if the injection pump unexpectedly shuts off. The check valve opening pressure should be at least 10 psi to prevent chemicals from draining from the supply tank into the irrigation system through the force of gravity.

Pressure Losses

These backflow prevention devices will cause pressure losses in the irrigation system. The greater the flowrate, the greater the pressure loss. One study found a pressure loss of less than 1 psi at a 1600 gallon-per-minute flowrate with the devices evaluated.

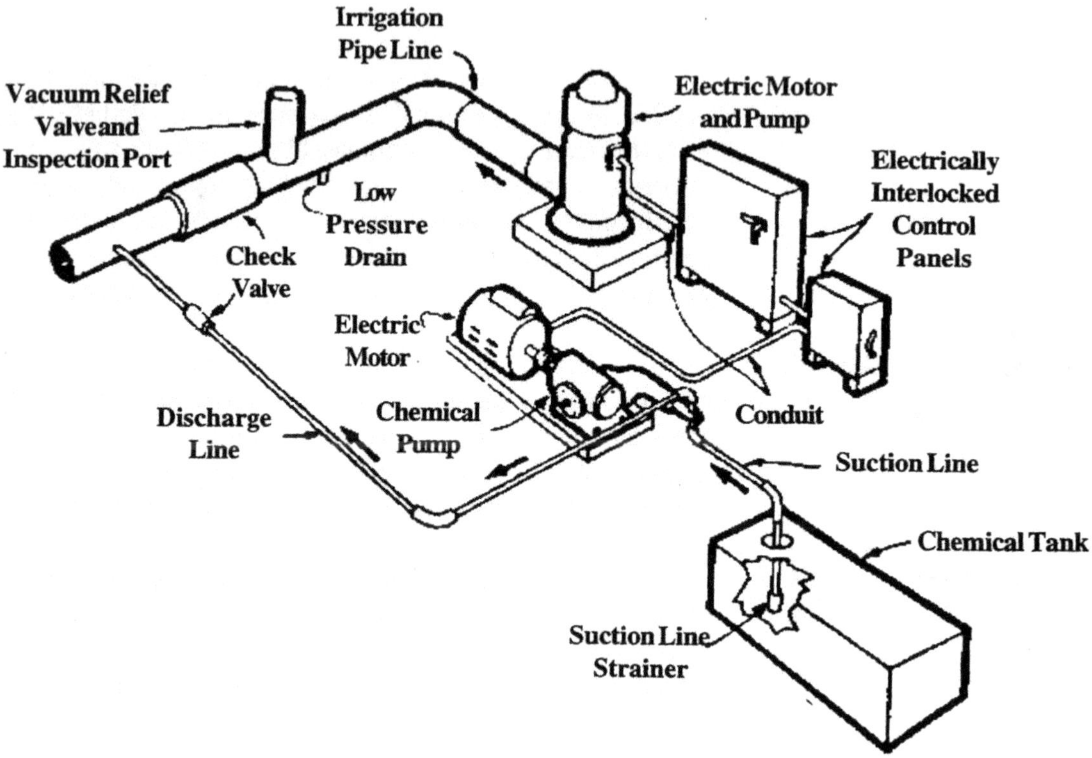

Figure 1. Components for backflow prevention.

References

Eisenhauer, D.E., H.E. Munir, and J.R. Gilley. 1988. Chemigation backflow prevention assemblies. *Proceedings of ASCE Conference, Planning Now for Irrigation and Drainage in the 21st Century,* Lincoln, Nebraska July 18-21, 1988.

Munir, H.M., D.E. Eisenhauer, and J.R. Gilley. 1992. Performance of chemigation backflow prevention assemblies. *Applied Engineering in Agriculture,* Vol. 8:211-19.

Trimmer, W., T.W. Ley, G. Clough, and D. Larsen. 1992. *Chemigation in the Pacific Northwest.* Pacific Northwest Extension Publication PNW360.

Soil Ingestion

By Blaine Hanson, UC Irrigation and Drainage Specialist

Soil ingestion can cause clogging problems with buried drip irrigation systems. Soil ingestion occurs when soil is sucked or ingested into the drip tape through the emitter openings. After the drip system is shut off, water in the laterals continues to drain down to the lower end of the field. The saturated soil immediately adjacent to the emitters, the result of the irrigation, prevents air from flowing into the lateral lines during drainage, and as a result a vacuum develops in the drip tape. This vacuum causes soil to be ingested into the drip line. The ingested soil can be deposited in the emitter flow passages, and plug the emitter. It also may be carried into the drip tape itself, and during the next irrigation, deposited in the lower end of the lateral.

What Affects Soil Ingestion

Little research exists on soil ingestion. However, field experience suggests that some factors affecting soil ingestion are the size of the emitter opening field slope, soil type, and length of lateral lines. Large emitter openings may allow more soil to be ingested than small openings. Steeper slopes and longer lateral lengths will tend to cause a larger vacuum in the drip tape than will smaller slopes and shorter lengths. Some soil types may move more readily into the emitter than other types.

Preventing Soil Ingestion

One approach to preventing soil ingestion is to install vacuum relief valves at the upslope end of the laterals, either on the water supply manifold or the flushing manifold. The vacuum relief valve will allow air to flow into the drip tape during drainage of the tape. The vacuum relief valve will prevent soil ingestion if the elevation of the manifold or submain is slightly higher than that of the drip tape. If the manifold elevation is less than that of the drip tape, then some water in the manifold or submain will be siphoned into the drip tape before the vacuum relief valve eliminates a vacuum in the drip tape.

Another approach is to use a drip tape that reduces or prevents soil ingestion. One manufacturer designed an emitter opening that closes if a vacuum occurs in the drip tape, thus reducing or preventing ingestion.

The Bottom Line

Will Using a Drip System Increase Profits?

By Blaine Hanson, UC Irrigation and Drainage Specialist

Using drip irrigation can increase profits by increasing crop yield, lowering water costs, and reducing costs for cultivation and other agricultural practices. On the minus side, however, drip irrigation can increase capital costs as much as $500 to $1300 per acre and can add additional costs for pressurizing and maintaining the irrigation system.

Whether converting to a drip system would be economically advantageous or not at a particular site depends to a large extent on how well the existing irrigation system is managed—specifically, whether the management of that system provides high yields along with an efficient use of water, leaving little to be gained by converting to a drip system.

Field data from Arizona on converting from furrow to drip irrigation in cotton found that drip irrigation increased yields and reduced the applied water by about one-half, but these results do not represent a valid basis for evaluating drip irrigation in California row crops because between 80 and 90 inches of applied water were used for the furrow irrigation in the Arizona fields—far more than is normally applied under furrow irrigation in California.

Comparing Yield Under Drip and Furrow Irrigation

Results in California have been mixed. Table 1 on the next page shows results from field-scale comparisons of drip irrigation and furrow irrigation of cotton in the San Joaquin Valley conducted by several agencies. The results generally show higher yields of cotton with the drip system, although several comparisons revealed nearly equal yields with the two systems. With one exception, less water was applied with the drip systems. Profits were generally higher for the furrow systems than for the drip systems. Higher yields would be needed to make the drip system more profitable.

California vegetable growers who have converted to drip irrigation report mixed economic results, with some experiencing yield increases and others experiencing no increase in yield. The available data is insufficient to draw conclusions about the reasons for these differences. At one location, a field-scale comparison of surface drip, subsurface drip, and furrow irrigation revealed little different in yield for three lettuce crops, but applied water ranged between 10 and 13 inches with the furrow system and between 5 and 8 inches with the drip system.

An analysis of the economics of drip irrigation prepared for the California Energy Commission, suggests the following:

• While drip irrigation uses less water than furrow irrigation, the reduced water costs alone are generally insufficient to offset the cost of converting to drip irrigation. If water savings alone are considered, water costs would have to amount to nearly $200 per acre-foot for drip irrigation to be more profitable than furrow irrigation.

• Drip irrigation is justified economically only if revenue is increased through higher yields on the irrigated land, through water saved to irrigate additional acreage, or through reduced fertilizer and cultural costs. Some growers have reported a savings in these costs of $100 to $200 per acre.

References

Boyle Engineering Corporation. 1994. Demonstration of Emerging Irrigation Technologies. Final Report to the State of California Department of Water Resources.

Constable, G.A. and A.S. Hodgeson. 1990. A comparison of drip and furrow irrigated cotton on a cracking clay soil 3. Yield and quality of four cultivar. *Irrigation Science*, Vol. 11:149-153.

Detar, W.R., G.J. Phene and D.A. Clark. 1992. Full cotton production with 24 inches of water. ASAE Paper 92-2607. Presented at the 1992 International Winter Meeting, Nashville, TN, December 15-18, 1992.

Fulton, A.E. 1995. *Subsurface drip irrigation: Eastern San Joaquin Valley.* Annual Report to the U.C. Salinity/Drainage Program and Prossier Trust.

Fulton, A.W., J.D. Oster, B.R. Hanson, C.J. Phene, and D.A. Goldhamer. 1991. Reducing drainwater: furrow vs. subsurface drip irrigation. *California Agriculture,* Vol. 45(2):4-8.

Hanson, B.R., L. J. Schwankl, K.F. Schulbach, and G. S. Pettygrove. 1997. A comparison of furrow, surface drip, and subsurface drip on lettuce yield and applied water. *Agricultural Water Management* (in press).

Hodgeson, A.S., G.A. Constable, G.R. Duddy, and I.G. Daniells. 1990. A comparison of drip and furrow irrigated cotton on a cracking clay soil 2. Water use efficiency, waterlogging, root distribution and soil structure. *Irrigation Science,* Vol. 11:143-148.

Mateas, L., J. Berengena, F. Organg, J. Diz, and E. Fereres. 1991. A comparison between drip and furrow irrigation in cotton at two levels of water supply. *Agricultural Water Management,* Vol. 19:313-324.

Table 1. Comparisons of furrow and drip irrigation.

Reference	Crop	Yield *		Water (inches)		Profit ($/ac)	
		Drip	Furrow	Drip	Furrow	Drip	Furrow
Fulton et al (1991)	Cotton	1,409	1,263	21	24	504	401
Mateos et al (1991)	Cotton	1,551	1,361	20	28	195	470
Hodgson et al (1990) and Constable & Dodgerson (1990)	Cotton	1,616	1,572	21	18	543	673
Boyle Engineering Corp (1994)							
1989	Cotton	1,527	1,081	23	29	465	247
1990	Cotton	1,290	1,274	24	20	279	437
1992	Cotton	1,566	1,365	23	20	493	507
1993	Cotton	1,465	1,295	18	25	440	429
Detar et al (1992)							
1990 (Good)	Cotton	1,704	1,738	24	42	596	684
1991 (Good)	Cotton	1,613	1,607	26	38	516	614
1990 (Poor)	Cotton	1,637	1,445	23	46	550	458
1991 (Poor)	Cotton	1,517	1,325	26	41	444	386
Hanson et al (1996)							
1991	Lettuce	18.6	19.6	4	10	-	-
1992	Lettuce	18.2	18.3	9	13	-	-
Fulton (1995)							
Variety 1	Tomato	47.7	45.3	27	38	-	-
Variety 2	Tomato	41.6	41.6	27	38	-	-

*Units used for yields are: cotton—pounds/ac; lettuce—tons/ac; tomato—tons/ac

Will Converting to Drip Irrigation Save Energy?

By Blaine Hanson, UC Irrigation and Drainage Specialist

Whether converting to drip irrigation will save energy depends partly on the conditions in the particular field and partly on what irrigation method was used before.

Converting from a sprinkler system to a drip system will usually result in an energy savings because the pressure requirements of the drip system are likely to be less than that of the sprinkler system. Converting to drip irrigation from a furrow irrigation system, on the other hand, may or may not save energy. The reason is that although the drip system will make it possible for less water to be applied to the field, thereby decreasing energy use, the drip system will also have a higher pressure requirement than the furrow system and will therefore require more energy. Energy will be saved, then, only if the energy saving from the decreased water need is greater than the increased energy needed to meet the drip system pressure requirements.

The energy use of different irrigation systems can be easily compared. Energy use is measured in either kilowatt-hours (Kwhr) or horsepower-hours, with kilowatts or horsepower representing the rate at which energy is used and hours representing the length of the operating time during which energy is consumed. The horsepower or kilowatts needed for an irrigation system depends on the pumping lift (how high the water must be raised), on the pressure requirements of the system, and on the system flowrate or capacity. The horsepower requirement can be calculated from the following equation:

$$HP = H \times Q \div (3960 \times E) \qquad (1)$$

Where HP = horsepower
H = feet of total head
Q = gallons per minute of pump capacity
E = efficiency of pump or pumping plant

The total head is the pumping lift added to the pump discharge pressure head (pressure head = 2.31 x pressure (psi).

Efficiency refers to either the overall efficiency of the pumping plant —including the efficiency of the motor or engine—or the efficiency of only the pump. Input horsepower is calculated using the overall efficiency, while brake horsepower—the shaft horsepower of the motor or engine—is calculated using only the pump efficiency

The pumping lift is determined by how far beneath the ground surface the water lies—which in turn depends on the characteristics of the water-bearing material or aquifer, on the design and construction of the well, and on the capacity of the pump.

The pump discharge pressure is determined by the pressure requirements of the emitters; by the amount of pressure lost in the laterals, manifolds, submains, mainlines, and valves because of friction; by pressure drops across filters; and by elevation differences in the field.

Comparing Energy Use

To compare the energy use of two different irrigation systems, use the following equations and follow the steps set out below:

$$\text{Kwhr/acre-foot} = 1.023 \times H \div E \qquad (2)$$

$$\text{Kwhr} = \text{Kwhr/acre-foot} \times \text{acre-feet of water} \qquad (3)$$

1. Calculate the energy use per acre-foot of the original irrigation system using *Equation 2.* The efficiency figure used should represent the overall pumping plant efficiency derived from a pumping plant test. For a surface system, the total head (H) will represent only the pumping lift, since the pressure requirements of conventional surface irrigation systems are minimal.

2. Use *Equation 3* to calculate total energy use, giving the volume of water applied to the field in acre-feet and the number of acres irrigated.

3. Repeat the procedure for the drip system. The total head of the drip system must include the discharge pressure head, which is calculated by multiplying the pressure in pounds per square inch by 2.31.

Example: A field-scale comparison of furrow and drip irrigation on a lettuce crop shows that 10.1 inches of water is applied with the furrow system , while 4.7 inches of water is applied with the drip system. Calculate the energy used by both systems where pumping lift is 100 feet, discharge pressure for the drip system is 40 psi, field size is 40 acres, and overall pumping plant efficiency is 55 percent.

Result:

1. The energy use of the furrow system:

$$\text{Kwhr/acre-foot} = 1.023 \times 100 \text{ feet} \div 0.55 = 186$$

Kwhr = 186 kwhr/acre-foot x 10.1 inches x 40 acres ÷ 12
= 6260

2. The energy use of the drip system:

Kwhr/acre-foot = 1.023 x (100 feet + 40 psi x 2.31) ÷ 0.55
= 1.023 x 192 ÷ 0.55
= 357

Kwhr = 357 kwhr/acre-foot x 4.7 inches x 40 acres ÷ 12
= 5590

In this example, overall energy use was lower for the drip system than for the furrow system—even though the kwhr per acre-foot of water pumped was much higher for the drip system than for the furrow—because the smaller amount of water applied by the drip system more than offset the greater pressure requirements of the drip system. The example illustrates that converting to drip irrigation from conventional surface irrigation will save energy only if the amount of water applied with the drip system is substantially less than that applied with the surface irrigation system.

Time-of-Use Electric Rates

Time-of-use electric rates (TOU) can reduce energy costs for irrigation. TOU rates consist of on-peak and off-peak periods although a partial-peak period sometimes is used. Higher electric rates are charged during the on-peak period, while low rates are charged during the off-peak period. The off-peak period is normally from 6 p.m. to noon the following day, and the on-peak period is usually from noon to 6 p.m.

New drip systems should be designed to operate only during the off-peak period. A design flowrate should be used that allows the field to be irrigated in 18 hours per day instead of 24 hours per day. Higher flowrates will be required, and slightly more pressure will be needed. Input horsepower will also be higher compared to operating 24 hours per day. The design flowrate can be calculated using:

$$Q = (449 \times A \times D)/T$$

where Q = design flowrate (gallons per minute), A = acres irrigated, D = inches of applied water, and T = hours of irrigation.

Example. Calculate the design flowrate for a 40-acre drip system to be irrigated daily during the period of peak evapotranspiration. The daily application is 0.35 inches per day.

Q = (449 x 40 acres x 0.35 inches per day)/(18 hours)
= 349 gallons per minute.

(The design flowrate for a 24-hour per day irrigation is 262 gallons per minute.)

Care should be used in converting an existing drip system to TOU operation. One approach is to simply reduce the daily irrigation time. This, however, may cause deficit irrigation during periods of peak evapotranspiration. Deficit irrigation can only be avoided if the system flowrate is sufficient to apply the needed water in 18 hours instead of 24 hours.

The flowrate into the field may need to be increased for TOU operation to apply the same volume of water in 18 hours as was applied in 24 hours. Increasing the flowrate may require using a drip tape with a higher tape discharge rate. Pressure and input horsepower will also increase because of the higher flowrate.

Appendix A

Glossary

Acidification. Adding sulfuric acid, phosphoric acid, or hydrochloric acid to a drip irrigation system in order to lower the water pH to counteract chemical precipitation, which can cause clogging.

Backwashing. Cleaning filters by reversing the flow of water or discharging water over the screen to wash away contaminants.

Batch tank. *See* ***Differential pressure tank.***

Chelation. Adding a polyphosphate such as sodium hexametaphosphate to the water to prevent iron from oxidizing.

Chemical precipitation. Process whereby chemicals present in water separate from solution and become deposited in the irrigation system, which can clog drip emitters.

Chlorination. Adding chlorine to an irrigation system to counteract clogging caused by organic materials such as algae, moss, and bacterial slime.

Coefficient of manufacturing variation (CV). A measure of the differences in output among emitters resulting from the manufacturing process.

Crop water use. The amount of water used by a crop in a specified period of time. *See also* ***Evapotranspiration.***

Deep percolation. Irrigation water percolating below the root zone where it is lost to crop production.

Differential pressure tank. Tank used for injecting chemicals through an irrigation system. The tank inlet is connected to the irrigation system at a point of pressure higher than that of the outlet connection, causing irrigation water to flow through the tank containing the chemical to be injected.

Discharge rate. The rate at which water is discharged from emitters, expressed as a volume per unit of time—gallons per hour or liters per hour.

Distribution uniformity (DU). *See* ***Emission uniformity.***

Double-chambered tape/tubing Drip tape or tubing consisting of a large main chamber, which conveys the water along the lateral, and a small secondary chamber with orifices in the outer wall through which water is discharged into the soil.

Efficiency. *See* ***Irrigation Efficiency.***

Emission uniformity. A measure of how uniformly water is applied over a field, calculated as the minimum depth of applied water (usually the average discharge of the lowest 25% of all emitters measured), divided by the average discharge of all emitters measured, multiplied by 100. Sometimes referred to as ***distribution uniformity.***

Emitter. Device through which water is discharged to the soil surface from the lateral lines in a drip system.

Emitter discharge coefficient. The relationship between the dimensions and configurations of flow passages and consequent pressure loss through the emitter.

Emitter discharge exponent. The relative ability of an emitter to compensate for variations in pressure. The smaller the exponent, the less sensitive the emitter discharge rate to pressure variation.

Evapotranspiration. The amount of water used in a specified period of time by a particular type of crop, consisting of water evaporating from the soil and water transpiring from the plants. Crop evapotranspiration estimates are available as either historical averages or real-time estimates from the California Department of Water Resources CIMIS program and the University of California Cooperative Extension offices.

Fertigation. Applying liquid fertilizer through an irrigation system.

Filtration. The removal of suspended solids from irrigation water.

Flowmeter. Device for measuring the flowrate of a drip irrigation system.

Friction loss. A loss of pressure caused by friction as water moves through the irrigation system.

Gypsum block. *See* ***Resistance block.***

In-line emitters. Drip emitters installed inside the drip tubing as part of the tubing flow path.

Irrigation efficiency. A measure of the amount of irrigation water beneficially used—principally for frost protection, leaching salts, and replenishing soil moisture—divided by the amount of water applied.

Lateral line. The water delivery pipeline or polyethylene hose that supplies water to the emitters from the mainlines or submains.

Leaching. Applying irrigation water in excess of the soil moisture depletion level to remove salts from the root zone.

Line-source. Drip tape or tubing with emission points at close regular intervals.

Mainlines and submains. The water delivery pipelines that supply water from the control head to the laterals.

Media filter. A filter consisting of a tank filled with filtering media such as silica sand or crushed granite that removes suspended solids from irrigation water by passing the water through the filtering media.

Net irrigation amount. The amount of water used by the crop since the last irrigation.

pH. A measure of the acidity of alkalinity of a liquid. A pH of 7.0 is neutral; a pH less than 7.0 is acidic; and a pH greater than 7.0 is alkaline.

Positive displacement pump. Piston or diaphragm pumps, powered by electricity, water, or gasoline, for injecting chemicals through an irrigation system at precise rates.

Propeller Flowmeter. Device consisting of a propeller linked by a cable or shafts and gears to a flow indicator for measuring flowrates in pipelines.

Resistance block. A device, usually made of gypsum, that measures the conductivity of current passing between two electrodes to indirectly measure soil moisture.

Salinity. Soil condition in which the salt concentration in the crop root zone is too high for optimum plant growth.

Sand Separator. Filter used to remove large, heavy particles from water.

Single-chambered tape/tubing. Drip tape or tubing consisting of a single main chamber. Water flows from this chamber into drip emitters spaced along the tape or tubing and is discharged from the emitters into the soil.

Soil amendment. A substance added to the soil primarily to improve the soil's physical condition.

Strip emitter. Type of emitter suitable for use with single-chambered tape or tubing. Strip emitters are long and narrow in shape and are usually classified as turbulent-flow emitters because the configuration of the flow path creates turbulence in the water, allowing for larger flow passages, which can reduce the potential for clogging.

Surface runoff. Water flowing off the lower end of the field.

Uniformity. *See* ***Emission uniformity.***

Venturi Device. Device for injecting chemicals through an irrigation system. The device consists of a constriction in the pipe flow area, resulting in a negative pressure or suction at the throat of the constriction.

Water amendment. A chemical added to water to improve soil-water properties, such as water infiltration.

Welded emitters. Individual emitters—usually turbulent flow emitters—that are welded to the side of the tape.

Appendix B

Calculating Friction Losses

By Blaine Hanson, UC Irrigation and Drainage Specialist

Knowing how much pressure is lost through the effect of friction in the irrigation system is important in making emitter discharge rates as uniform as possible.

Pressure losses in pipes and hoses are commonly described in "feet of head" (H_f). A column of water 2.31 feet high will create a pressure of 1 psi at its base.

Pressure losses in drip tape, tubing, submains, manifolds, and main-lines can be calculated using any one of several equations, depending in part on the type of pipe material used. Pressure losses in plastic pipes, hoses, tubing, and tape can be calculated by the following equation:

$$H_f = 1050 \div D^{4.87} \times (Q \div C)^{1.852} \qquad (1)$$

Where: H_f = feet of pressure lost per 100 feet of pipe length
Q = flowrate in gallons per minute
D = internal pipe diameter in inches
C = friction coefficient (depending on pipe material)

Pressure loss can also be described in psi per 100 feet by dividing H_f by 2.31 (1 psi = 2.31 feet of water).

Table 1 on the following page gives values for C.

Table 1. Friction coefficients (C) for various pipe materials.

pipe material	*friction coefficient (C)*
aluminum	130
steel (new)	130
steel (15 years old)	100
concrete	100
cement asbestos	140
galvanized steel	135
drip tape/tubing	135
plastic pipe (4" or greater)	150
plastic pipe (2" to 3")	140

• ***Pressure losses along pipeline length.*** Where the flowrate is constant along the pipeline, pressure losses from friction along the length of a pipeline can be calculated by multiplying H_f by the number of 100-foot sections along the pipeline (found by dividing the pipeline length by 100).

• ***Adjusting for multiple outlets.*** Where the flowrate gradually decreases along the pipe length because water is being discharged through more than one outlet — as through emitters along a lateral or several laterals along a manifold — the figure representing pressure loss along the pipeline length must be multiplied by a reduction coefficient *(F)* as shown in *Table 2*, below to obtain an approximate estimate of actual friction loss.

Table 2. Pressure loss reduction coefficients (F).

Number of outlets	*Reduction coefficient (F)*
1	1.00
2	0.52
3	0.44
4	0.41
5	0.40
6	0.39
7	0.38
8	0.38
9	0.37
10-11	0.37
12-15	0.37
16-20	0.36
21-30	0.36
>31	0.36

Calculate the pressure loss along a 600-foot lateral in a drip tape with a diameter of 0.625 inches. The inlet flowrate is 2.5 gpm and the emitters are spaced 12 inches apart.

Step 1. Calculate H_f using *Equation 1:*

$$H_f = (1050 \div 0.625^{4.87}) \times (2.5 \div 135)^{1.852} = 6.4 \text{ feet}$$

Step 2. Calculate the pressure loss from friction for the lateral length (H_L):

$$H_L = 6.4 \text{ feet} \times 6 = 37.2 \text{ feet}$$

Step 3. Adjust H_L for 600 outlets. From *Table 2, F* = 0.36.

$$H_L = 0.36 \times 37.2 \text{ feet} = 13.4 \text{ feet}$$

Step 4. Convert to psi:

$$\text{psi} = 13.4 \text{ feet} \div 2.31 \text{ feet/psi} = 5.8 \text{ psi}$$

Reference

Keller, J., and R.D. Bliesner. 1990. *Sprinkle and trickle irrigation.* Van Nostrand and Reinhold, New York.

Appendix C

Designing Laterals for Drip Irrigation Systems

By Blaine Hanson, UC Irrigation and Drainage Specialist

There are several expensive and sophisticated design programs available for sizing laterals for drip irrigation systems. These programs generally are not for grower use. A simple procedure is presented to help size lateral lengths. Information needed is the design lateral inlet pressure (depends on tape thickness and type of material), the tape discharge characteristics, and the design emission uniformity. A hand calculator with a "log" key and an exponent key (y^x) is needed for this procedure.

Procedure

A. Determine the allowable pressure loss for the design emission uniformity ($EU_{Lateral\ Pressure}$) due to pressure variation.

Tape Discharge Characteristics

1. Determine the tape discharge characteristics (discharge exponent and discharge coefficient). Information needed are the tape discharge rates (Q_1, Q_2) at two different pressures (P_1 P_2). This information must be obtained from the manufacturer or the Center for Irrigation Technology.

$$\text{Discharge Exponent (X)} = \log (Q_1/Q_2) / \log (P_1/P_2)$$

$$\text{Discharge Coefficient (K)} = Q_2 / P_2^X$$

2. Calculate the tape discharge rate (Q_I) at the lateral inlet for the inlet design pressure (P_I).

$$Q_I = (K)(P_I^X)$$

3. Calculate the tape discharge rate (Q_E) at the end of the lateral length for the design emission uniformity ($EU_{Lateral\ Pressure}$).

$$Q_E = (21.6)(Q_I) / (118.8 - EU_{Lateral\ Pressure})$$

4. Calculate the pressure (P_E) at the end of the lateral.

$$P_E = (Q_E / K)^{1/X}$$

Allowable Pressure Loss

5. Calculate the allowable pressure loss (P_A) and head loss (H_A) for the design emission uniformity.

$$P_A = P_I - P_E$$

$$H_A = (P_A)(2.31)$$

Lateral Length

B. Determine the lateral length that will result in a pressure or head loss about equal to the allowable pressure loss (P_A) or head loss (H_A). Either existing computer design programs or the following trial-and-error procedure can be use.

1. Calculate the average tape discharge rate.

$$Q_A = (Q_I + Q_E) / 2$$

2. Assume a lateral length (L) and calculate the lateral inflow rate (Q_L).

$$Q_L = (Q_A)(L)/100$$

3. Calculate the head loss due to friction and slope (S). D is the internal diameter (inches) of the drip tape.

$$H_L = (0.00048)(Q_L^{1.75})(L) / D^{4.75} - (S)(L)$$

(S = feet of fall per foot of distance or slope expressed as a percent divided by 100, I. E. 0.1%/100. If the flow is uphill, add (S)(L). If the flow is downhill, subtract (S)(L)).

4. Repeat steps 2 and 3 until a H_L is about equal to H_A.

Lateral Uniformity

C. Calculate the $EU_{Lateral}$ due to both pressure variation and coefficient of manufacturing (CV).

$$EU_{Lateral} = (100)(EU_{Lateral\ Pressure})(EU_{CV})$$

where $EU_{CV} = 100(1 - 1.27 \times CV)$

(EU_{CV} and $EU_{Lateral\ Pressure}$ are expressed as decimals for calculating EU.)

Steep Slopes

This procedure gives reasonable results for uphill flow, for flat conditions, and for downhill slopes less than about 0.5%. Under these conditions, the minimum tape discharge rate is at the end of the lateral. For large downhill slopes, the minimum tape discharge rate may not occur at the end of the lateral. This trial-and-error procedure could underestimate the design lateral length.

Example

Determine an appropriate lateral length for a high-flow drip tape. Use the following design data.

$CV = 0.026$

$EU_{Lateral\ Pressure} = 95\%$

$P = 10$ psi
Slope = 0.2% (downhill)
Tape diameter = 5/8" (0.625")

$Q_1 = 0.36$ gpm/100feet, $P_1 = 8$ psi
$Q_2 = 0.54$ gpm/100 feet, $P_2 = 14$ psi

A. Determine allowable pressure or head loss for $EU_{\text{Lateral Pressure}}$.

1. Determine tape discharge characteristics.

$$X = \log(0.36/0.54) / \log(8/14) = 0.72$$

$$K = 0.54 / 14^{0.72} = 0.081$$

2. Calculate the tape discharge rate at the lateral inlet for the design pressure.

$$Q_I = (0.081)(10^{0.72}) = 0.43 \text{ gpm/100feet}$$

3. Calculate the tape discharge rate at the end of the lateral.

$$Q_E = (21.6)(0.43) / (118.8 - 95) = 0.39 \text{ gpm/100 ft}$$

4. Calculate the pressure at the end of the lateral.

$$P_E = (0.39 / 0.081)^{1/0.72} = 8.9 \text{ psi}$$

5. Calculate the allowable pressure loss and head loss.

$$P_A = 10 - 8.9 = 1.1 \text{ psi}$$

$$H_A = (1.1)(2.31) = 2.5 \text{ feet}$$

B. Determine the appropriate lateral length either with a computer design program or by trial-and-error.

1. Calculate the average tape discharge rate.

$$Q_A = (0.43 + 0.39) / 2 = 0.41 \text{ gpm/100 feet}$$

2. Assume a lateral length and calculate the lateral inflow rate.

$L = 400$ feet
$$Q_L = (0.41)(400) / 100 = 1.64 \text{ gpm}$$

3. Calculate the head loss for 400 feet.

$$H_L = (0.00048)(1.64^{1.75})(400) / 0.625^{4.75} - (0.002)(400)$$
$$= 3.4 \text{ feet}$$

4. Since H is slightly greater than H , assume a new lateral length and repeat steps 2 and 3. If a new length of 350 feet is assumed, then H = 1.4 feet, which is less than H_A. Thus, a lateral length between 350 feet and 400 feet will provide an $EU_{LATERAL\ PRESSURE}$ of about 95%.

C. Determine the $EU_{Lateral}$ for the lateral.

$$EU_{CV} = 100(1 - 1.27 \times 0.026 / 1.414) = 98\%$$

$$EU_{Lateral} = 100(0.98)(0.95) = 93\%$$

Appendix D

How Much Water Does the Crop Need?

By Blaine Hanson, UC Irrigation and Drainage Specialist

Table 1. Estimated seasonal water requirements for California row crops by location*.

Region	Crop	Season	Water requirement (acre-inches/acre)
Northeastern Mountain Valleys	Corn	June-September	24
	Potatoes	January-August	22
North Coast Interior Valleys	Corn	June-September	22
Central Coast Coastal Valley	Beans (dry)	June-August*	11
	Carrots	July-October*	14
	Cauliflower	August-December	16
	Corn	June-September	17
	Lettuce	February-May*	12
	Lettuce	May-July	15
	Lettuce	July-September*	11
	Potatoes	June-October*	17
	Sugarbeets	March-October	27
	Tomatoes (canning)	May-September	19
Central Coast Interior Valleys	Beans (dry)	June-August	14
	Corn (field)	May-September	22
	Sugarbeets	March-October	34
	Tomatoes (canning)	May-September	24
South Coast Interior Valleys	Potatoes	June-August	19
	Sugarbeets	March-October	34
South Coast Coastal Valleys and Plains	Strawberries	August-April*	25
	Tomatoes (canning)	March-September	25
Sacramento Valley	Beans	Apr. 30-Aug. 18	20
		June 1-Sept. 21	20
	Corn (field)	Apr. 30-Aug. 25	27
		May 1-Sept. 12	23
		June 15-Oct. 19	27

	Potatoes	May-July	20
	Sorghum	May 13-Sept. 29	23
		June 17-Oct. 20	19
		July 1-Oct. 31	16
	Sugarbeets	June-December*	30
	Tomatoes	Feb. 26-Aug. 11	25
		May 1-Sept. 22	28
		June 3-Sept. 29	22
San Joaquin Valley	Beans (dry)	Apr. 1-July 31	21
		May 1-Aug. 15	21
		June 15-Sept. 30	18
	Cantaloupes	April-July	13
	Carrots (seed)	mid Jan/early Feb-mid-late July	23
	Corn (field)	Mar. 15-Aug. 15	27
		Apr. 1-Sept. 15	28
		Apr. 15-Sept. 15	27
		May 15-Sept. 30	24
		June 15-Oct. 15	22
	Cotton	Apr. 1-Sept. 30	31
		Apr. 16-Oct 15	32
		May 1-Oct. 31	32
	Melons	Feb 15-June 30	16
		Mar. 15-July 31	21
		Apr. 15-Aug. 15	19
	Onions	Mar. 1-Sept. 30	32
		Sept. 15-May 31	20
		Nov. 15-July 31	34
	Potatoes	Jan. 1-May 15	13
		Feb. 1-June 15	17
		Mar. 1-June 30	20
	Sugarbeets	Feb. 1-Aug. 31	37
		June 15-Mar. 15	29
		Mar. 15-Sept. 15	35
		May 1-Jan. 31	36
	Tomatoes	Apr. 1-Sept. 30	26
		May 1-Sept. 15	25
Southeastern Desert	Asparagus	January-December	67
	Cantaloupe	February-May	23
	Carrots	October-April	26
	Cotton	April-October	42
	Lettuce	Sept.-Dec.	8
	Squash	Sept.-Dec.	11
	Sugarbeets	July-April	46
	Tomatoes (canning)	February-June	37

Appendix E

COST PER POUND OF NUTRIENT ELEMENT

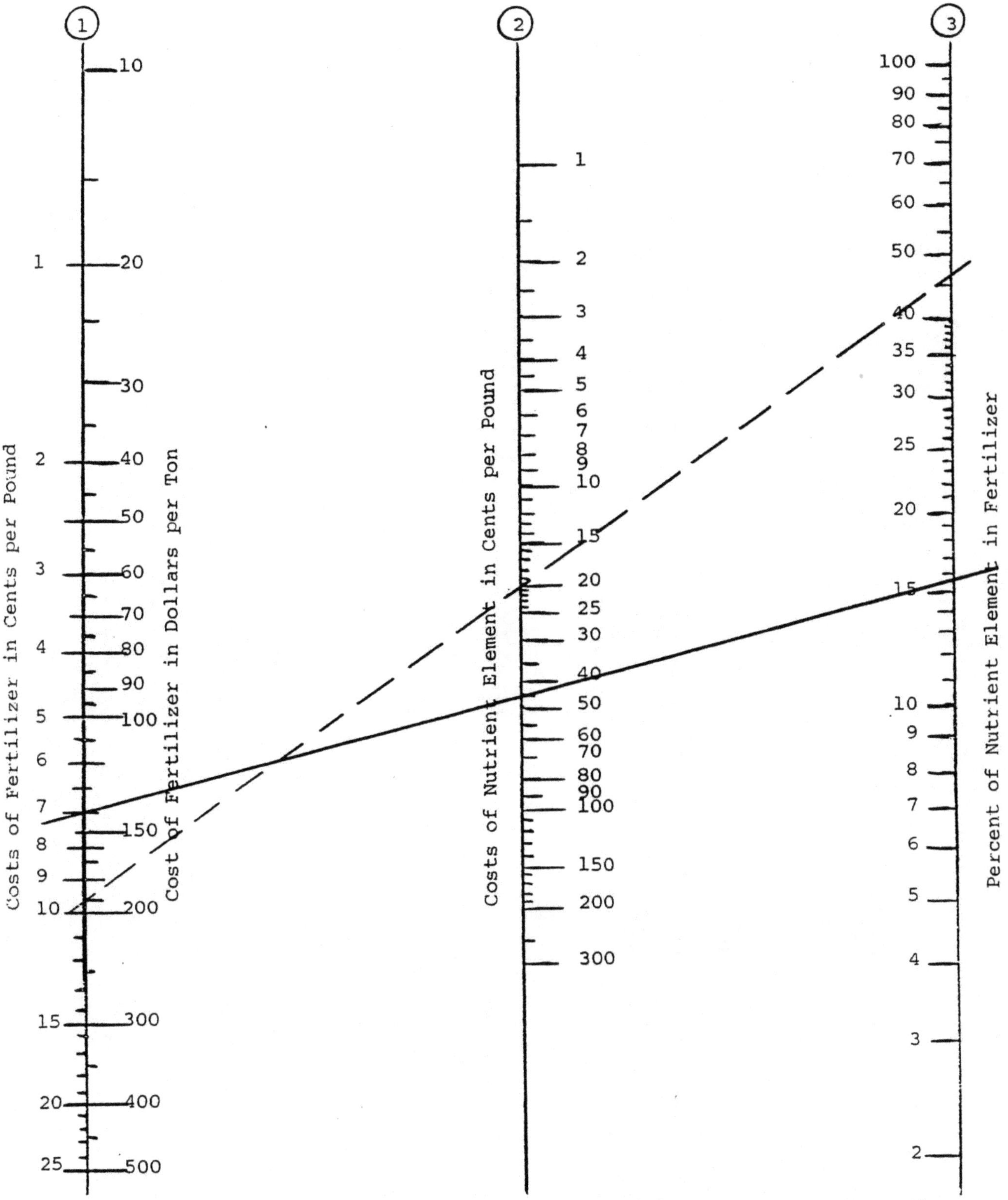

To calculate the cost per pound of nutrient element, it is necessary to know the cost per ton (or per pound) of the fertilizer to be used and the percent of nutrient element in the fertilizer.

Example: Compare cost per pound of nitrogen for 46% urea at $190 per ton and 15.5% calcium nitrate at $140 per ton.

First - Locate $190 per ton on Scale 1 and 46% on Scale 3. Connect these values. Answer read on Scale 2 is 21.7 cents per pound.

Second - Locate $140 per ton on Scale 1 and 15.5 on Scale 3. Connect these values. Answer read on Scale 2 is 45.1 cents per pound.

Index

O

P

R

S

T

U

CPSIA information can be obtained
at www.ICGtesting.com
Printed in the USA
FSHW011210310319
56780FS

9 781601 079220